PREPPER SURVIVAL FOR BEGINNER

"Mastering Disaster Readiness: A Comprehensive Guide for Beginners and Experts, Unveiling Survival Skills, Security Strategies, and Life-Saving Techniques for All Scenarios."

Tim m. Holland

All rights reserved. No part of this publication may be reproduced, distributed, or transmitted in any form or by any means, including photocopying, recording, or other electronic or mechanical methods, without the prior written permission of the publisher, except in the case of brief quotations embodied in critical reviews and certain other noncommercial uses permitted by copyright law.

Copyright © Tim m. Holland , 2024.

TABLE OF CONTENTS

Introduction: Unveiling Your Path to Preparedness

Why Prepping Matters in 2024: A World in Flux

From Beginner to Expert: A Roadmap for Building Confidence

Dispelling Myths and Common Misconceptions

Part 1: Building a Solid Foundation (2024

Chapter 1: Assessing Threats and Tailoring Your Plan (The 2024 Landscape)

Evaluating Your Lifestyle and Vulnerability Factors

Setting Realistic and Achievable Goals for Preparedness (2024 Focus)

Chapter 2: The Essential Emergency Kit: Your 72-Hour Lifeline

Customised for Specific Needs

Chapter 3 : Mastering Food and Water Security: A Sustainable Approach (2024 Focus)

Water Purification Techniques: From Simple to Advanced

Chapter 4: Shelter Solutions: From Staying Put to Evacuating Safely
Shelter Solutions: From Staying Put to Evacuating Safely

Adapting Your Strategy for Different Threats (Fire, Flood, Flood, Power Outages)

Chapter 5: Firecraft and Cooking Without Utilities: Mastering the Basics

Firecraft and Cooking Without Utilities: Mastering the Basics (continued)

From Foraged Feast to Food Poisoning: A Cautionary Tale (For Experienced Users Only)

Chapter 6: First Aid and Basic Medical Care: Be Ready to Respond

First Aid and Basic Medical Care: Be Ready to Respond

Chapter 7: Navigation and Communication When Technology Fails: Finding Your Way Without GPS

Navigation and Communication When Technology Fails: Sending the Message When Lines Are Down

Chapter 8: Security and Self-Defence Strategies: Protecting Yourself and Your Loved Ones

Part 3: Mastering Long-Term Preparedness

Chapter 9: Staying Informed and Monitoring Threats: Knowledge is Power

Understanding Local Emergency Alert Systems

Building Strong Community Networks for Mutual Support

Chapter 10: Mental and Physical Preparedness: Building Resilience - Managing Stress and Anxiety in Uncertain Times

Maintaining Physical Fitness and Overall Health: The Bedrock of Preparedness

Cultivating a Positive Mindset and Adaptability: Essential Tools for Preparedness

Chapter 11: Sharpening Your Skills and Refining Your Plan - Conducting Regular Drills for Evacuation Scenarios and First Aid

Maintaining and Upgrading Your Prepping Supplies: Ensuring Readiness Over Time

Continuously Learning New Skills to Stay Ahead of the Curve: A Pillar of Preparedness

Conclusion: The Empowered Prepper: Ready for Anything - The Importance of Long-Term Planning and Staying Vigilant

Peace of Mind Through Preparedness: A Lifelong Journey

Bonus Appendix:

Master Prepper Checklist: A Comprehensive List of Essential Supplies (Printable)

Glossary of Prepping Terms: Understanding the Language of Preparedness

Sample Emergency Plans: Customizable Templates for Family and Evacuation Planning

Introduction: Unveiling Your Path to Preparedness

The world around us is constantly changing. Natural disasters, economic upheavals, and unforeseen events can disrupt our lives in an instant. While we can't control the future, we can take steps to be prepared for whatever may come our way. Prepping isn't about paranoia or living in fear; it's about empowerment and taking control of your own safety and security.

This book is your guide to becoming a prepper, regardless of your experience level. Here, we'll unveil a clear path to preparedness, one that starts with understanding the **"why"** behind prepping and evolves into the **"how"** of building a solid foundation for survival.

Why Prepare in 2024?

The year 2024 may hold particular challenges. Perhaps you've witnessed extreme weather events in your region or felt the sting of economic uncertainty. Maybe you simply want to feel more confident in your ability to weather any storm. Whatever your reasons, being prepared brings peace of mind and allows you to focus on helping yourself and your loved ones during difficult times.

From Beginner to Expert: A Roadmap to Confidence

This book is designed to be your companion on your prepping journey. Whether you're a complete novice or looking to refine your existing skills, you'll find valuable information here. We'll start by addressing common misconceptions about prepping and establish a realistic approach that fits your lifestyle and needs.

This introduction sets the stage for the chapters to come, which will delve into:

- **Building a strong foundation:** Assessing potential threats, assembling essential supplies, and creating a sustainable food and water plan.
- **Unveiling essential survival skills:** Mastering firecraft, first aid, navigation, and communication without relying on technology.
- **Mastering long-term preparedness:** Maintaining your physical and mental well-being, staying informed, and refining your plan over time.

Why Prepping Matters in 2024: A World in Flux

The year 2024 seems to hum with a subtle undercurrent of uncertainty. Perhaps it's the memory of recent extreme weather events - record-breaking heat waves, devastating floods, or powerful storms - that have left a mark. Maybe it's the whispers of economic instability or the ever-present possibility of unexpected disruptions to our daily routines. Whatever the reason, a sense of unease lingers for many.

In this climate of flux, prepping isn't just a good idea; it's a proactive step towards securing your well-being and the well-being of your loved ones. Here's why prepping matters in 2024:

- **Heightened Risk of Disasters:** Climate change seems to be intensifying, with a greater frequency and severity of natural disasters.

From wildfires scorching vast landscapes to hurricanes ripping through coastal communities, these events can leave communities devastated and infrastructure crippled. Having a plan and essential supplies in place can make a world of difference in these situations.

- **Potential Supply Chain Disruptions:** The globalised world we live in can be vulnerable to disruptions in international trade. Political tensions, economic downturns, or even cyberattacks could lead to shortages of essential goods - from food and medicine to fuel and basic necessities. A well-stocked pantry and understanding alternative means of obtaining resources can provide a safety net during such times.

- **Economic Uncertainty:** The economic landscape can be unpredictable, and periods of recession or high inflation can strain household budgets. Having a stockpile of non-perishable food and other essentials can help you weather these storms with less financial stress.

- **Empowerment and Self-Sufficiency:** Prepping isn't about living in fear; it's about taking responsibility for your own safety and security. By acquiring essential skills and building a preparedness plan, you empower yourself to handle unexpected situations with confidence.

This self-reliance can be a source of immense comfort during challenging times.

Prepping in 2024 isn't about creating a doomsday bunker; it's about taking practical steps towards peace of mind. By being prepared for a range of potential disruptions, you can navigate challenges with greater resilience and ensure the well-being of yourself and those you care about. The next chapter will delve into dispelling common myths about prepping, paving the way for a realistic and empowering approach to preparedness.

From Beginner to Expert: A Roadmap for Building Confidence

The world of prepping can seem daunting, especially for beginners. Images of heavily fortified bunkers and survivalists living off the grid might come to mind. However, prepping is a journey, not a destination.

This book will be your roadmap, guiding you from your first steps as a novice to a place of confidence and preparedness, regardless of your current experience level.

Here's how we'll build your confidence on this journey:

1. Small Steps, Big Impact: We won't overwhelm you with complex tasks. We'll start with small, achievable goals, like assembling a basic emergency kit or learning how to purify water. As you master these steps, your confidence will grow, motivating you to tackle more advanced skills.

2. Focus on Practicality: This book prioritises practical skills applicable to real-world situations. We'll focus on building a preparedness plan that fits your specific needs and lifestyle. You won't find instructions on building elaborate traps or unrealistic survival scenarios.

3. Breakdown Complex Skills: We'll break down complex skills like first aid or navigation into manageable steps. Clear instructions, diagrams, and even potential practice exercises will equip you with the knowledge to handle emergencies with confidence.

4. Celebrate Your Progress: We'll encourage you to celebrate your progress, no matter how small. As you build your emergency kit, learn to build a fire, or refine your evacuation plan, acknowledge your accomplishments. These accomplishments will keep you motivated and on the path.

5. A Supportive Community: Prepping doesn't have to be a solitary pursuit. We'll discuss ways to connect with other preppers in your community, online forums, or local preparedness groups. Sharing experiences and knowledge with others can bolster your confidence and provide valuable support on your prepping journey.

Dispelling Myths and Common Misconceptions

The world of prepping is often shrouded in misconceptions. Sensationalised media portrayals and Hollywood depictions can create a distorted view of what prepping truly is. Let's debunk some of the most common myths and set the record straight:

Myth #1: Preppers are Crazy Doomsdayers:

This stereotype paints preppers as fanatics obsessed with the apocalypse. In reality, most preppers are simply ordinary people who want to be prepared for common emergencies like power outages, severe weather events, or unexpected supply chain disruptions. They value self-reliance and want to ensure their families' well-being during challenging times.

Myth #2: Prepping Requires a Bunker and Tons of Money:

While some preppers do have elaborate setups, that's not the norm. Prepping can be a gradual process, starting with a basic emergency kit and building your supplies over time. You don't need a fortune to get started - focus on acquiring essential items and building skills that are practical for your location and potential threats.

Myth #3: Prepping is About Stockpiling Guns and Ammo:

While some preppers may choose to invest in firearms for self-defence, that's not a central focus for everyone. Prepping is more about building resilience, acquiring essential skills, and having a plan in place to handle a variety of situations.

Myth #4: Prepping is Only for Remote Locations:

Regardless of where you live, disruptions to daily life can occur. Power outages, floods, or even local infrastructure issues can impact anyone. Having a basic preparedness plan is beneficial for urban dwellers just as much as those in rural areas.

Myth #5: Prepping is All About Survival:

Prepping is more than just about surviving a worst-case scenario. It's about being prepared to handle disruptions and inconveniences with minimal stress. A well-stocked pantry and a plan can ensure your family can weather a power outage or a temporary job loss without significant hardship.

By dispelling these myths, we can see prepping for what it truly is: a proactive approach to managing risk and ensuring your well-being. The next chapter will delve into the first essential step of your prepping journey - assessing threats and tailoring your plan.

Part 1: Building a Solid Foundation (2024 Focus)

The year 2024 presents a unique opportunity to establish a strong base for your goals and aspirations. This first part will focus on the essential elements that will act as the bedrock for your success throughout the year.

1. Self-Assessment:

Begin by taking stock of your current situation. Analyse your strengths, weaknesses, opportunities, and threats (SWOT analysis). Identify what worked well for you in the past year, and acknowledge areas that need improvement. Consider emerging trends and potential challenges specific to 2024.

2. Goal Setting:

Once you have a clear understanding of your current position, it's time to set SMART goals (Specific, Measurable, Achievable, Relevant, and Time-bound). Be ambitious yet realistic, considering the resources and timeframe available. Break down large goals into smaller, actionable steps to maintain momentum.

3. Building Core Skills:

The ever-evolving world demands continuous learning. Identify in-demand skills relevant to your field and personal development goals. Explore online courses, attend workshops, or consider pursuing certifications to enhance your skill set and stay competitive in 2024.

4. Prioritisation and Time Management:

With a clear set of goals, prioritise tasks effectively. Utilise time management tools and techniques to ensure you dedicate sufficient time and energy to high-impact activities.

5. Building a Support System:

Surround yourself with positive and encouraging individuals. Seek mentors who can offer guidance and share their experiences. Build a network of colleagues or friends who can provide support, accountability, and celebrate your achievements.

A Focus on 2024:

As you build your foundation in 2024, stay informed about current events, technological advancements, and economic trends that might influence your goals. This awareness will allow you to adapt your strategies and make informed decisions throughout the year.

By following these steps and maintaining a focus on continuous improvement, you can establish a solid foundation that will empower you to achieve success in 2024 and beyond.

Chapter 1: Assessing Threats and Tailoring Your Plan (The 2024 Landscape)

Identifying Potential Disasters in Your Region

A crucial aspect of building a strong foundation for 2024 is understanding the potential threats you might face. This chapter focuses on identifying disasters specific to your region, allowing you to tailor your plan for increased resilience.

Understanding Your Location:

- **Research common disasters:** Start by researching the most frequent and impactful disasters in Lagos, Nigeria.

Reliable sources include government websites (National Emergency Management Agency - NEMA), news archives, and academic journals

- **Consider local factors:** Lagos, a coastal megacity, faces unique vulnerabilities. Flooding due to heavy rains and clogged drainage systems is a major concern . Additionally, densely populated areas are susceptible to fire outbreaks.
- **Emerging threats:** Be aware of potential new threats. Climate change might increase the frequency and intensity of floods and storms. Rapid urbanisation could exacerbate existing vulnerabilities.

Common Disaster Categories:

While the specific threats will vary by location, some general disaster categories to consider include:

- **Natural Disasters:** Floods, droughts, landslides, earthquakes (though less common in Lagos)
- **Man-Made Disasters:** Industrial accidents, hazardous material spills, fire outbreaks

Assessing Your Vulnerability:

Once you have identified potential disasters, evaluate your personal vulnerability:

- **Location:** Do you live in a low-lying area prone to flooding? Is your neighbourhood densely built, increasing fire risk?
- **Housing:** Is your building structurally sound? Does it have proper fire safety measures?

- **Infrastructure:** Is your community equipped with functional drainage systems and emergency response protocols?

Tailoring Your Plan:

By understanding potential threats and your vulnerability, you can tailor your plan for increased preparedness:

- **Develop a disaster preparedness kit:** Assemble essential supplies like food, water, first-aid kits, and communication tools to sustain you during a disaster.
- **Create an evacuation plan:** Identify safe evacuation routes and designated meeting points for your family in case of emergencies.
- **Stay informed:** Monitor weather forecasts and emergency alerts. Register for local notification systems to receive timely warnings.

Evaluating Your Lifestyle and Vulnerability Factors

Building a solid foundation for 2024 goes beyond simply setting goals. It's crucial to understand how your lifestyle choices and inherent vulnerabilities can impact your ability to achieve those goals. This self-assessment will empower you to identify areas for improvement and tailor your plan for success.

Lifestyle Choices:

- **Health Habits:** Evaluate your diet, sleep patterns, and exercise routine. Are you making healthy choices that will support your physical and mental well-being throughout the year? Consider areas for improvement – prioritising nutritious meals, incorporating regular exercise, and aiming for adequate sleep.

- **Financial Habits:** Assess your spending habits and financial security. Do you have a budget in place? Are you saving for your goals? Identify areas where you can cut unnecessary expenses and develop strategies to build a financial safety net.
- **Time Management:** Analyse how you spend your time. Are you constantly feeling overwhelmed and behind schedule? Explore time management techniques and identify areas where you can streamline your routine and dedicate focused time to your priorities.
- **Stress Management:** Evaluate your stress levels and coping mechanisms. Do you have healthy outlets for managing stress? Consider relaxation techniques like meditation or yoga to maintain a sense of balance and well-being throughout the year.

Vulnerability Factors:

Beyond lifestyle choices, consider inherent vulnerabilities that might impact your plans:

- **Health Conditions:** Do you have any pre-existing health conditions that require ongoing management? Factor in potential limitations and necessary adjustments to your goals and plan accordingly.
- **Social Support System:** Evaluate the strength of your social network. Do you have reliable friends and family who can offer support and encouragement throughout the year? A strong support system can significantly enhance your resilience and motivation.
- **Access to Resources:** Consider your access to essential resources like education, healthcare, and technology.

Limited access might require creative problem-solving and identifying alternative approaches to achieving your goals.

Actionable Steps:

By critically evaluating your lifestyle and vulnerabilities, you can create a more realistic and achievable plan for 2024:

- **Set SMART goals:** Considering your lifestyle and vulnerabilities, refine your goals to be Specific, Measurable, Achievable, Relevant, and Time-bound.
- **Develop contingency plans:** Identify potential roadblocks related to your lifestyle or vulnerabilities and brainstorm alternative strategies to overcome them.

- **Seek support:** If you identify areas where your lifestyle or vulnerabilities need improvement, seek support from professionals, friends, or online resources.

Setting Realistic and Achievable Goals for Preparedness (2024 Focus)

Disasters can strike unexpectedly, disrupting our lives and throwing our plans into disarray. However, by setting realistic and achievable preparedness goals, you can significantly increase your resilience and navigate challenges with greater confidence.

The Power of SMART Goals:

When setting preparedness goals, utilise the SMART framework:

- **Specific:** Clearly define the desired outcome of your goal. Instead of a vague goal like "be prepared," aim for something specific like "assemble a 3-day emergency kit by June 1st."
- **Measurable:** Establish a way to track your progress. For your emergency kit goal, you could measure progress by listing down the items you need and checking them off as you acquire them.
- **Achievable:** Be honest about your resources and limitations. Don't overwhelm yourself with an overly ambitious goal. Building a comprehensive emergency kit might take time, so start with smaller, achievable steps like gathering non-perishable food items over a few weeks.

- **Relevant:** Ensure your goals align with the potential threats you identified in Chapter 1. If flooding is a major concern, focus on building a kit that includes waterproof items and flotation devices.
- **Time-bound:** Set a deadline for achieving your goal. By doing this, you maintain focus and generate a sense of urgency.
- **Examples of SMART Preparedness Goals for 2024:**
- **Financial Preparedness:** Allocate a specific amount of your monthly budget towards building an emergency fund.
- **Communication Plan:** By July 1st, identify a designated contact person outside your region and establish a communication plan for emergencies.

- **Home Safety:** Schedule a professional inspection of your home's electrical wiring and plumbing by the end of June.
- **Basic Skills:** Enrol in a first-aid and CPR certification course offered in your community by August.

Additional Tips for Setting Realistic Preparedness Goals:

- **Start Small:** Don't try to do everything at once. Begin with small, manageable goals and gradually build upon them.
- **Focus on Progress:** Celebrate your achievements, no matter how small. This will motivate you to stay on track.

- **Be Flexible:** Life throws curveballs. Be prepared to adjust your goals as needed based on unforeseen circumstances.

The Takeaway:

Setting realistic and achievable preparedness goals empowers you to take control and build a stronger foundation for navigating the challenges of 2024. By following the SMART framework and focusing on progress, you can create a personalised preparedness plan that fosters peace of mind and allows you to focus on achieving your long-term goals.

Chapter 2: The Essential Emergency Kit: Your 72-Hour Lifeline

Disasters can disrupt access to basic necessities. An emergency kit, stocked with essential supplies, can be your lifeline during the critical first 72 hours following a disaster. This chapter will guide you through assembling a comprehensive emergency kit, ensuring you have the core supplies for survival: food, water, shelter, first aid, and communication.

Core Supplies for Your 72-Hour Kit:

1. Food:

- Non-perishable food: Pack enough non-perishable, ready-to-eat meals and snacks to sustain each person in your household for 3 days (72 hours).

Consider high-calorie, energy-dense options like canned goods, protein bars, and dried fruits.

- **Manual can opener:** A can opener is essential for accessing canned food if power outages disable electric can openers.

2. Water:

- **Water:** Store one gallon of water per person per day for at least three days. This includes water for drinking and sanitation.

  Consider purchasing commercially bottled water or investing in water purification tablets or a portable water filter in case of contamination.

3. Shelter:

- **Emergency blanket:** These lightweight, reflective blankets can help retain body heat in cold weather.

- **Poncho or raincoat:** Protect yourself from rain and wind with a waterproof poncho or raincoat. Consider a multi-purpose emergency shelter if available in your area.

4. **First Aid:**

- **First-aid kit:** Assemble a first-aid kit stocked with essential supplies like bandages, antiseptic wipes, pain relievers, and any prescription medications your family requires.

5. **Communication:**

- **Battery-powered or hand-crank radio:** Stay informed about weather updates and emergency broadcasts during power outages. A NOAA weather radio with a tone alert is ideal.
- **Extra batteries:** Ensure you have enough batteries to power your radio and flashlights.

- **Cell phone charger:** A portable phone charger will allow you to maintain communication during power outages. Consider a solar-powered charger for long-term use.

Additional Considerations:

- **Flashlight:** Include a flashlight with extra batteries for nighttime emergencies. Headlamps can be particularly useful for providing hands-free lighting.
- **Hygiene items:** Pack essential hygiene items like toilet paper, hand sanitizer, toothbrush, and toothpaste for basic sanitation.
- Important documents: Include copies of important documents like passports, insurance cards, and identification in a waterproof container.

- **Cash:** Keep some cash in small denominations on hand in case electronic transactions are unavailable.
- Multi-tool: A multi-tool can be a valuable resource for various tasks.

Maintaining Your Kit:

- Regularly check and replace expired items.
- Rotate food and water supplies to maintain freshness.
- Store your kit in a cool, dry, and easily accessible location.

Customised for Specific Needs

Disasters can disrupt access to basic necessities, impacting everyone differently. While a core emergency kit provides a foundation, customising it for specific needs within your household is crucial. This chapter explores tailoring your kit for individuals with medical conditions, children, and pets.

Core Supplies for Your 72-Hour Kit (Refer to Chapter 1 for details):

- Food (non-perishable)
- Water
- Shelter (emergency blanket, poncho)
- First-aid kit
- Communication (battery-powered radio, extra batteries, cell phone charger)
- Flashlight (with extra batteries)
- Hygiene items (toilet paper, hand sanitizer, etc.)

- Important documents (copies in waterproof container)
- Cash (small denominations)
- Multi-tool

Customising Your Kit:

1. Medical Conditions:

- **Medications:** Include a sufficient supply of essential prescription medications for at least 72 hours for each family member. Consider alternative administration methods if electricity is unavailable.
- **Medical Supplies:** People with specific medical conditions may require additional supplies, such as syringes, catheters, or diabetic testing kits. Consult with your healthcare provider to determine necessary supplies.

- **Medical Information:** Include a laminated medical information sheet listing each person's allergies, medications, and contact information for their healthcare provider.

2. Children:

In addition to core supplies:

- **Food and Water:** Pack familiar, non-perishable foods and snacks children enjoy to minimise stress during emergencies.
- **Comfort Items:** Include comforting items like stuffed animals, favourite blankets, or books to provide a sense of security for children.
- **Entertainment:** Pack age-appropriate games, puzzles, or colouring books to keep children occupied during stressful situations.

- **Identification:** Consider wristbands or ID tags for young children with contact information for parents or guardians.

3. Pets:

In addition to core supplies:

- **Food and Water:** Include a 3-day supply of non-perishable pet food and water in bowls or collapsible containers.
- **Leash, Harness, and Waste Bags:** Ensure you have a leash or harness to control your pet and waste bags for proper sanitation.
- **Medications:** If your pet requires any medications, include a sufficient supply in your kit.
- **Pet Carrier:** A sturdy pet carrier will provide a safe haven for your pet during an evacuation.

- **Veterinarian Contact Information:** Keep a copy of your veterinarian's contact information in your emergency kit.

Long-Term Storage and Maintenance

While assembling a comprehensive emergency kit is crucial, its effectiveness hinges on proper storage and regular maintenance. This section will guide you on selecting the ideal storage location and maintaining your kit for optimal functionality throughout the year.

Choosing the Right Location:

- **Cool and Dry:** Store your kit in a cool, dry, and climate-controlled location. Avoid attics, basements, or garages prone to extreme temperature fluctuations or moisture build-up, which can damage food, water, and medications.

- **Easily Accessible:** Select a location readily accessible during an emergency. Avoid storing your kit in locked areas or behind heavy objects. Family members, particularly children and elderly individuals, should be able to locate and access the kit with ease.
- **Elevated from the Floor:** Store your kit on a shelf or elevated platform to protect it from potential flooding or water damage.

Essential Storage Containers:

- **Durable and Waterproof:** Invest in sturdy, airtight containers to shield your kit's contents from dust, moisture, and pests. Waterproof containers offer added protection from accidental spills or leaks.

- **Labelling:** Clearly label each container with its contents and the date of last inspection. This facilitates easy identification and retrieval during an emergency.

Maintaining Your Emergency Kit:

- **Regular Inspections:** Conduct thorough inspections of your kit at least twice a year, preferably before the start of peak disaster seasons in your region.
- **Food and Water Rotation:** Rotate food and water supplies to ensure freshness. Consume older items before adding new ones to your kit.
- **Replace Expired Items:** Discard and replace any expired food, water, medications, or batteries.

- **Update Information:** Review and update any critical information stored in your kit, such as medical information sheets or contact details.

Benefits of Proper Storage and Maintenance:

By prioritising proper storage and maintenance, you can ensure your emergency kit remains effective and ready for use whenever needed. This proactive approach fosters peace of mind and empowers you to respond confidently to unexpected disasters.

Chapter 3 : Mastering Food and Water Security: A Sustainable Approach (2024 Focus)

Disasters can disrupt access to fresh food and clean water, jeopardising your health and well-being. This chapter focuses on building a sustainable, non-perishable food stockpile that caters to various dietary needs, ensuring food and water security during emergencies.

The Importance of Non-Perishable Food Stockpiles:

A well-stocked pantry with non-perishable food items forms the cornerstone of your emergency preparedness plan. This stockpile allows you to maintain a sense of normalcy and provides essential sustenance during the critical first 72 hours following a disaster.

Building a Sustainable Stockpile:

While stocking up on non-perishable food is crucial, a sustainable approach considers several factors in 2024:

- **Nutritional Value:** Aim for a variety of non-perishable foods that provide essential nutrients like carbohydrates, protein, and healthy fats. Consider incorporating options with extended shelf life that are also minimally processed and have limited added sugars or sodium (a growing health concern in 2024).
- **Shelf Life:** Prioritise items with extended shelf lives, minimising waste and ensuring a readily available food source during emergencies. Opt for canned goods stored in BPA-free cans whenever possible to minimise potential health risks.

- **Dietary Needs:** Cater to specific dietary requirements within your household. Consider options for vegetarians, vegans, individuals with allergies, or those with religious dietary restrictions.
- **Personal Preferences:** Include familiar and enjoyable non-perishable foods to maintain morale and encourage consumption during stressful times.

Non-Perishable Food Options for Different Diets:

Here's a breakdown of non-perishable food options suitable for various dietary needs:

- **General Stockpile:**

    - Canned fruits and vegetables (in BPA-free cans if possible) with reduced added sugars (consider options packed in water)
    - Canned beans and lentils (high in protein and fibre)
    - Canned meats (tuna, chicken, salmon) in water or broth (avoid options with high sodium content)
    - Dried fruits and nuts (excellent source of energy and healthy fats)
    - Whole-grain crackers and brown rice (long shelf life and source of complex carbohydrates)

- Natural nut butters (high-protein option)
- Shelf-stable milk (carton or powdered) with lower sugar content (if dairy is tolerated)
- Granola and protein bars with limited added sugars (convenient and energy-dense)

- **Vegetarian/Vegan Options:**

  - Canned beans and lentils (staple source of protein)
  - Canned chickpeas and tofu (versatile protein options)
  - Dried fruits, nuts, and seeds (nutrient-rich and high in energy)
  - Whole-grain pasta and brown rice (long shelf life and source of complex carbohydrates)

- Canned vegetarian chili or soup with lower sodium content
- Vegetable broth (adds flavour and nutrients to meals)
- Vegan protein bars and trail mix with limited added sugars

- **Considerations for Allergies:**

  - Choose gluten-free options like canned beans, rice, nuts, and dried fruits if gluten intolerance exists.
  - Opt for dairy-free alternatives like almond milk or shelf-stable soy milk for lactose intolerance.
  - Carefully review ingredient labels to avoid potential allergens.

Additional Tips for Building a Sustainable Stockpile in 2024:

- **Buy in Bulk (when practical):** Purchasing staples like whole-grain brown rice or dried beans in bulk can be cost-effective for those with ample storage space.
- **Rotate Stock:** Regularly rotate your stockpile by consuming older items and replacing them with fresh ones to prevent spoilage.
- **Consider Food Preparation:** Include basic cooking tools like a manual can opener and a camping stove (with proper fuel storage) to prepare meals during emergencies when electricity might be unavailable.

Water Security:

Water is vital for survival. In addition to your emergency kit's water supply, consider long-term water security options:

- **Bottled Water:** Stockpile bottled water with extended shelf lives, prioritising BPA-free options whenever possible.
- **Water Purification Tablets/Filters:** Invest in water purification tablets or a portable water filter to treat potentially contaminated water sources.

Water Purification Techniques: From Simple to Advanced

Access to clean water is essential for survival. Disasters can disrupt water treatment facilities, leaving you with potentially contaminated water sources. This chapter explores various water purification techniques, empowering you to make informed decisions and ensure safe drinking water during emergencies.

The Importance of Water Purification:

Contaminated water can harbour harmful bacteria, viruses, and parasites that can cause serious illness. Water purification eliminates or reduces these contaminants, making water safe for consumption.

Simple Water Purification Techniques:

These methods require minimal resources and are suitable for short-term emergencies:

- **Boiling:** The simplest and most effective method. Bring water to a rolling boil for at least one minute (at higher altitudes, boil for a longer duration) to kill most pathogens. Let the water cool completely before consumption.
- **Chlorination:** Add chlorine tablets or bleach (following manufacturer's instructions) to disinfect water. This method requires waiting time for the chlorine to take effect. Important: Never mix bleach with ammonia, as this creates toxic fumes.

Moderate Water Purification Techniques:

These methods offer a higher level of filtration and are suitable for longer-term use:

- Sedimentation: Allow suspended particles to settle at the bottom of a container for several hours. Carefully decant the clear liquid from the top, leaving sediment behind. This improves water clarity but does not eliminate all contaminants.
- **Filtration:** Use a portable water filter to remove impurities and pathogens. These filters come in various types, with different pore sizes and filtration capabilities. Choose a filter certified to remove bacteria, viruses, and protozoa.

Advanced Water Purification Techniques:

These methods are ideal for long-term preparedness or situations with highly contaminated water sources:

- **Distillation:** Boils water and condenses the steam, leaving contaminants behind. Distillation units are effective but require a heat source and can be slow.
- **Reverse Osmosis:** Uses a semi-permeable membrane to remove a wide range of contaminants, including dissolved salts and minerals. Reverse osmosis systems are complex and require a pressure source.

Choosing the Right Water Purification Technique:

The ideal water purification method depends on several factors:

- **Water Source:** Consider the level of contamination of your water source.
- **Availability of Resources:** Evaluate the resources available during an emergency, such as fuel for boiling or replacement filters.
- Desired Level of Filtration: Determine the level of filtration needed based on the potential contaminants in your water source.
- **Number of People:** Choose a method that can provide enough clean water for everyone in your household.

Additional Considerations:

- **Water Storage:** Store purified water in clean, airtight containers to prevent recontamination.
- **Regular Maintenance:** For certain purification methods, like portable filters, follow proper cleaning and maintenance procedures to ensure effectiveness.

Chapter 4: Shelter Solutions: From Staying Put to Evacuating Safely

Disasters can damage homes, jeopardising your safety and security. This chapter explores various shelter solutions, empowering you to make informed decisions and ensure your well-being during emergencies.

The Importance of Shelter Solutions:

A safe and secure shelter protects you from the elements, hazards, and potential dangers associated with disasters. This chapter focuses on two primary shelter solutions: fortifying your home as a safe haven and planning for safe evacuation procedures.

Part 1: Fortifying Your Home as a Safe Haven (Emergency Sheltering, Basic Repairs)

Emergency Sheltering:

- **Identify a Safe Room:** Designate a safe room within your home, preferably an interior room on the lowest floor and away from exterior walls, windows, and doors. This room should be sturdy enough to withstand potential hazards.
- **Prepare Your Safe Room:** Stockpile emergency supplies in your safe room, including first-aid kits, non-perishable food, water, and a battery-powered radio. Consider reinforcing the door with a deadbolt lock for added security.
- **Window and Door Protection:** Board up windows with plywood or hurricane shutters to protect against flying debris during storms or high winds. Reinforce garage doors if necessary.

Basic Repairs:

- **Roof Maintenance:** Regularly inspect and repair your roof to minimise water damage during storms. Secure loose shingles and consider mitigation techniques like hurricane straps for high-wind areas.
- **Plumbing Shut-Off Valve:** Locate the shut-off valve for your water supply. Knowing its location allows you to quickly turn off the water in case of a pipe break or other plumbing emergencies.
- **Electrical Panel:** Familiarise yourself with the location of your electrical panel. Knowing how to shut off the electricity in an emergency can prevent electrical fires.

Part 2: Evacuating Safely (Planning and Procedures)

Evacuation Planning:

- **Develop an Evacuation Plan:** Create a comprehensive evacuation plan for your household, identifying potential evacuation routes, designated meeting locations, and communication strategies.
- **Stay Informed:** Monitor weather reports and emergency alerts during disaster threats.

  Observe the local authorities' directions for evacuation.

  Practise Your Evacuation Plan: Regularly rehearse your evacuation plan with all family members, ensuring everyone understands their roles and responsibilities.

Evacuation Procedures:

- **Assemble Your Emergency Kit:** Grab your pre-assembled emergency kit before evacuating.
- **Turn Off Utilities:** If time permits, turn off utilities like gas, water, and electricity at the main shutoff valves or panels to minimise the risk of further damage.
- **Leave Quickly and Safely:** Evacuate promptly using designated evacuation routes. Avoid taking unnecessary risks or attempting to retrieve belongings once an evacuation order is issued.

Additional Considerations:

- **Shelter for Pets:** Include a plan for evacuating or sheltering pets during emergencies. Consider pet-friendly evacuation shelters or create a designated safe space for pets within your home if staying put.
- **Individuals with Disabilities:** Develop specific evacuation procedures considering the needs of individuals with disabilities within your household. Ensure they have access to assistance and appropriate transportation during evacuation.

Shelter Solutions: From Staying Put to Evacuating Safely

Disasters can damage homes, jeopardising your safety and security. This chapter explores various shelter solutions, empowering you to make informed decisions and ensure your well-being during emergencies.

The Importance of Shelter Solutions:

A safe and secure shelter protects you from the elements, hazards, and potential dangers associated with disasters. This chapter focuses on two primary shelter solutions: fortifying your home as a safe haven and planning for safe evacuation procedures.

Part 1: Fortifying Your Home as a Safe Haven (Emergency Sheltering, Basic Repairs)

Refer to previous sections on fortifying your home as a safe haven and basic repairs.

Part 2: Evacuating Safely (Planning and Procedures)

Evacuation Planning:

- **Develop an Evacuation Plan:** Create a comprehensive evacuation plan for your household, identifying potential evacuation routes, designated meeting locations, and communication strategies.
- **Stay Informed:** Monitor weather reports and emergency alerts during disaster threats. Heed evacuation orders issued by local authorities.

- **Practise Your Evacuation Plan:** Regularly rehearse your evacuation plan with all family members, ensuring everyone understands their roles and responsibilities.

Evacuation Planning: Bug-Out Bag Essentials

Your Bug-Out Bag (BOB) is a portable survival kit designed to sustain you for the initial phase of an evacuation, typically 72 hours. Here are essential items to include:

- **Basic Supplies:**

    - Non-perishable food (3-day supply)
    - Water (1 gallon per person per day)
    - First-aid kit
    - Medications (prescription and over-the-counter)

- Sanitation items (toilet paper, wipes, hygiene products)
- Flashlight with extra batteries
- Cash (small bills)
- Multipurpose tool
- Whistle or emergency signal mirror
- Maps of the local area and potential evacuation routes
- Copies of important documents (ID, insurance) in a waterproof container
- Rain gear and sturdy shoes
- Cell phone charger (portable option if possible)

- **Additional Considerations:**

  o Include specific items for infants, children, or elderly individuals within your household.
  o Pack pet supplies if you plan to evacuate with your pets (food, water, leash, waste bags, medications).
  o Consider adding a change of clothes, a comfort item (blanket for children), and entertainment (books, games) for extended evacuations.

Choosing an Evacuation Destination:

- **Identify Potential Shelters:** Research potential evacuation shelters located along designated evacuation routes or outside your local disaster zone. Consider shelters with pet-friendly policies if necessary.
- **Stay with Family or Friends:** If possible, designate out-of-town family or friends as alternative evacuation destinations, offering a familiar and potentially less crowded environment.
- **Communication Plan:** Ensure your evacuation plan includes a designated out-of-area contact person to facilitate communication with loved ones after evacuating.

Additional Considerations:

- **Shelter for Pets:** Include a plan for evacuating or sheltering pets during emergencies. Consider pet-friendly evacuation shelters or create a designated safe space for pets within your home if staying put.
- **Individuals with Disabilities:** Develop specific evacuation procedures considering the needs of individuals with disabilities within your household. Ensure they have access to assistance and appropriate transportation during evacuation.

Adapting Your Strategy for Different Threats (Fire, Flood, Flood, Power Outages)

Disasters come in many forms, each requiring a tailored preparedness approach. This chapter guides you on adapting your emergency plan and shelter strategy for specific threats: fire, flood, and power outages.

Understanding the Threat:

- **Fire:** Rapidly spreading flames and smoke pose the primary dangers. Evacuation is often crucial.
- **Flood:** Rising water levels can cause property damage, disrupt infrastructure, and create hazardous conditions. Sheltering in place or evacuation on higher ground might be necessary.
- **Power Outages:** Loss of electricity can disrupt essential services and creature comforts. While generally less urgent than other disasters, extended outages can impact daily life.

Adapting Your Emergency Plan:

- **Fire:**

    o **Focus on Evacuation:** Practice evacuation drills regularly, ensuring everyone in your household knows escape routes and meeting locations.

    o **Install Smoke Alarms:** Equip your home with smoke alarms on every floor and outside sleeping areas. Test them monthly and replace batteries regularly.

    o **Create a Fire Escape Plan:** Plan multiple escape routes from each room, considering alternative exits if windows are blocked by flames.

- **Flood:**

  - **Monitor Flood Risks:** Identify your flood risk zone and stay informed about potential flooding during heavy rains or storms.
  - **Develop a Flood Evacuation Plan:** Plan evacuation routes considering potential road closures. Identify higher ground refugees within your neighbourhood or designated evacuation shelters outside the flood zone.
  - **Flood-Proofing Measures:** Consider sandbags or other measures to protect doorways and potential water entry points if sheltering in place.

- **Power Outages:**

    - **Alternative Lighting:** Include flashlights, headlamps, or lanterns in your emergency kit to provide illumination during outages.
    - **Communication Plan:** Establish a communication plan in case phone lines are down. Consider a battery-powered radio or predetermined meeting locations.
    - **Food and Water Storage:** Maintain a stockpile of non-perishable food and water to sustain your household during extended outages.
    - **Alternative Heat Source:** If residing in a cold climate, consider alternative heating options like a fireplace or a camping stove (with proper ventilation and fuel storage) for warmth during outages.

Adapting Your Shelter Strategy:

- **Fire:** Evacuate your home immediately upon hearing a smoke alarm or noticing a fire. Do not waste time gathering belongings.
- **Flood:** Follow evacuation orders if instructed to do so by authorities. If sheltering in place, move to the highest level of your home and avoid flooded areas.
- **Power Outages:** Shelter in place unless extreme temperatures threaten your health.

  Wear layers so you can adapt to any temperature changes.

Part 2: Unveiling Essential Survival Skills

While a comprehensive emergency kit provides a foundation, preparedness extends beyond stockpiling supplies. Mastering essential survival skills empowers you to adapt to challenging situations and navigate emergencies with greater confidence. This part focuses on core skills crucial for various disaster scenarios.

1. Fire Safety and Escape:

- **Fire Prevention:** Practise fire safety measures like proper electrical wiring maintenance and avoiding overloading outlets. Keep combustible objects far from heat sources.

- **Fire Extinguisher Training:** Learn how to use a fire extinguisher properly for small fires.
- **Stop, Drop, and Roll:** Reinforce the "Stop, Drop, and Roll" technique for extinguishing clothing fires.

2. Basic First Aid:

- **CPR and First Aid Certification:** Consider obtaining CPR and basic first aid certification to handle injuries and medical emergencies.
- **Wound Care:** Learn wound cleaning, dressing, and bandaging techniques.
- **Bleeding Control:** Understand methods for controlling bleeding, including direct pressure and pressure points.

3. Search and Rescue:

- **Basic Search Techniques:** Develop basic search techniques for locating lost individuals, especially important for wilderness emergencies.
- **Signalling for Help:** Learn how to use whistles, signal mirrors, or bonfires to signal for help in remote locations.

4. Navigation:

- **Map and Compass Skills:** Master basic map and compass navigation skills to find your way even without GPS or cell phone reception.
- **Natural Navigation:** Learn to use natural signs like the sun and stars for basic directional orientation.

5. Shelter Building:

- **Temporary Shelters:** Develop skills for building basic temporary shelters using available materials like branches and tarps.
- **Emergency Insulation:** Understand methods for creating emergency insulation using natural materials or clothing to conserve body heat.

6. Signalling for Help:

- **International Distress Signals:** Learn international distress signals like SOS (three short bursts, three long bursts, three short bursts) to attract attention during emergencies.
- **Fire Signalling:** Understand how to build and maintain signal fires to alert rescuers.

7. Firecraft:

- **Building a Fire:** Master basic fire-starting techniques using matches, lighters, or primitive methods like flint and steel.
- **Fire Safety and Maintenance:** Learn how to safely build, maintain, and extinguish campfires.

8. Food and Water Procurement:

- **Safe Water Sources:** Identify safe water sources in your environment and understand basic water purification techniques like boiling or filtration.
- **Edible Plants:** Learn to identify edible plants and their safe consumption methods (consult a field guide and prioritise professional instruction before foraging).
- **Signalling for Help:** Understand basic trapping and fishing techniques to supplement your food supply in survival situations.

9. Weather Awareness and Response:

- **Understanding Weather Patterns:** Learn to identify basic weather patterns and interpret weather forecasts to anticipate potential threats.
- **Seeking Shelter During Storms:** Know where to seek safe shelter during different types of storms (e.g., tornadoes, hurricanes, blizzards).

10. Maintaining Positive Mental Attitude:

- **Stress Management Techniques:** Develop coping mechanisms to manage stress and maintain a positive mental attitude during challenging situations.
- **Signalling for Help:** Learn basic relaxation techniques like deep breathing exercises to manage anxiety and improve decision-making.

Chapter 5: Firecraft and Cooking Without Utilities: Mastering the Basics

Disasters can disrupt access to utilities, making the ability to build a fire a crucial survival skill. In addition to providing heat and light, fire can be used to cook food and purify water. This chapter equips you with the knowledge and techniques to build a fire under any condition, ensuring you can navigate emergencies with confidence.

The Importance of Firecraft:

- **Warmth and Comfort:** Fire provides essential warmth during cold weather emergencies, preventing hypothermia.
- **Light Source:** Fire offers illumination during power outages, improving visibility and safety at night.

- **Cooking and Water Purification:** Fire allows you to cook food safely and purify potentially contaminated water by boiling.
- **Signalling for Help:** A well-maintained fire can be used as a distress signal to attract rescuers in remote locations.

Building a Fire Under Any Condition:

The key to building a successful fire lies in proper preparation and understanding the fire-building process. Here's a breakdown of the essential steps:

1. **Gather Tinder:**
- **Selection:** Choose dry, easily combustible materials like tinder bundles, dry leaves, or wood shavings.
- **Preparation:** Prepare a generous amount of tinder to ensure a successful ignition. Fluffier tinder catches sparks more readily.

2. **Build Your Kindling:**
- **Break Twigs:** Snap dry twigs into finger-sized pieces to create kindling. These bridge the gap between tinder and larger fuelwood.
- **Arrange the Kindling:** Build a teepee-shaped structure with the kindling, leaving space in the centre for tinder.

3. **Fuelwood Selection:**
- **Size Matters:** Start with small, pencil-thick sticks and gradually progress to larger logs as the fire establishes itself.
- **Dry Wood is Key:** Use only seasoned, dry firewood for efficient burning. Damp wood produces smoke and struggles to ignite.

4. Lighting the Fire:

- **Spark Generation:** Use matches, a lighter, or a fire starter (flint and steel) to create sparks and ignite your tinder bundle.
- **Gently Blow:** Gently blow on the embers to encourage a steady flame.

5. Maintaining the Fire:

- **Feed the Fire Gradually:** Add progressively larger pieces of firewood as the fire grows, maintaining a manageable size.
- **Stack Logs Carefully:** Lean logs against the existing fire, allowing air circulation for optimal burning.

Building a Fire in Challenging Conditions:

- **Wet Conditions:** Gather dry tinder from under logs, overhangs, or inside dead trees. Shave dry wood from larger branches to create tinder.
- **Windy Conditions:** Build your fire in a sheltered location or create a wind barrier using rocks or logs. Light the fire on the upwind side to aid ignition.

Tinder Selection Tips:

- **Natural Tinder:** Dry leaves, bark, wood shavings, or shredded dry grass are all excellent natural tinder options.
- **Commercial Tinder:** Consider carrying commercially available tinder packs or tinder starters for convenience and reliability.

- **Cotton Balls with Petroleum Jelly:** Soak cotton balls in petroleum jelly to create a fire-starting tinder that is compact and weather-resistant.

Additional Tips:

- **Practice Fire Building**: Regularly practise building fires in a safe controlled environment to hone your skills and gain confidence.
- **Fire Safety Precautions:** Clear a safe area around your fire and never leave it unattended. Before departing the campsite, totally put out the fire.
- **Be Responsible:** Use firewood responsibly and avoid collecting live wood or stripping bark from trees.

By following these guidelines and practising firecraft techniques, you can ensure you have the ability to build a fire and harness its benefits whenever needed during an emergency

Firecraft and Cooking Without Utilities: Mastering the Basics (continued)

Building a fire is a valuable skill, but mastering safe and efficient cooking methods during emergencies extends beyond firecraft. This section explores alternative cooking options for various situations, ensuring you can prepare nutritious meals even when utilities are unavailable.

Safe and Efficient Cooking Methods:

1. **Camp Stoves:**
- Fuel Options: Camp stoves come in various fuel types like propane, butane, isobutane-propane (isopro), or white gas.

  Choose a fuel type readily available in your region and consider factors like portability and environmental impact.

- Safe Operation: Prioritise operating your camp stove outdoors with proper ventilation to avoid carbon monoxide poisoning. Follow manufacturer instructions for safe fuel handling, stove setup, and ignition.

- Cooking Versatility: Camp stoves offer a versatile cooking platform suitable for boiling water, frying, or simmering food. Pots and pans compatible with your stove are essential for meal preparation.

2. **Alternative Fuels:**

- Solid Fuels: Solid fuel tablets or compressed wood blocks are lightweight and compact options for short-term cooking needs. Always use these fuels in a designated camp stove or fire pit for safety.
- Alcohol Stoves: Simple and lightweight, alcohol stoves burn readily available denatured alcohol. However, they can be less efficient compared to other options and may require a windscreen for optimal performance in windy conditions.

3. Solar Cooking:

- Harnessing the Sun: Solar cookers utilise the power of the sun to cook food slowly and safely. They come in various designs, including parabolic cookers or box cookers. These options are best suited for long cooking times and require clear skies for optimal performance.

4. Fire Cooking:

- Cooking Over an Open Fire: While open fire cooking offers a traditional method, it requires practice and attention to detail.
- Utensils and Cooking Techniques: Utilise grilling tools like roasting forks, skewers, or cast iron cookware for fire cooking. Master techniques like pot hanging or food wrapping in foil packets for efficient cooking over an open flame.

Choosing the Right Cooking Method:

The most suitable cooking method depends on several factors:

- Emergency Situation: Consider the length of the power outage or disaster. Camp stoves or alternative fuels offer faster cooking compared to solar cookers.
- Available Resources: Choose a method compatible with the resources at hand. Opt for camp stoves if fuel is available, or utilise alternative fuels or fire cooking if resources are limited.
- Cooking Needs: Evaluate your cooking requirements. Do you need to boil water quickly, simmer stews, or simply heat pre-cooked meals? Choose a method that meets your specific needs.

Additional Considerations:

- Food Safety: Always prioritise food safety practices, particularly during emergencies. Maintain proper hygiene, cook food thoroughly, and store perishable items safely.
- Water Conservation: Utilise minimal water for cooking whenever possible. Consider reusing cooking water for other purposes like washing dishes.
- Campsite Etiquette: If using a fire for cooking, adhere to local fire restrictions and practise responsible campfire management.

From Foraged Feast to Food Poisoning: A Cautionary Tale (For Experienced Users Only)

WARNING: This section is intended for experienced foragers only. Improper identification of wild plants can lead to serious illness or even death. Only consume plants you can confidently identify through multiple reliable sources.

This chapter explores the advanced concept of utilising foraged resources for culinary creativity during emergencies. Remember, foraging should only be attempted by individuals with extensive knowledge of wild plants in their specific region.

The Allure of Foraged Feasts:

In dire circumstances, foraging can provide a valuable source of supplementary food. However, the risks associated with improper identification far outweigh the potential benefits.

For experienced foragers, wild plants offer a unique opportunity to add variety and nutritional value to their emergency meals.

Essential Foraging Principles:

- Positive Identification: Never consume a wild plant unless you can identify it with 100% certainty using multiple reliable field guides and cross-referencing with online resources from reputable botanical organisations.
- Start Slowly and Simply: Begin by foraging for a few easily recognizable species and gradually expand your repertoire as your confidence grows.
- Respect the Environment: Practise sustainable harvesting techniques. Only take what you need, avoid disturbing plant roots, and leave enough for future growth and wildlife consumption.

Incorporating Foraged Ingredients:

- Leaves and Greens: Tender leaves of certain wild plants can be utilised in salads, stir-fries, or cooked as potherbs. Examples include dandelion greens, chickweed, or wood sorrel (consult references for safe consumption guidelines).
- Fruits and Nuts: Wild berries and nuts can add sweetness and nutritional variety to your diet. Research proper ripening times and avoid consuming any unknown fruits or nuts.
- Roots and Tubers: Experienced foragers can explore identifying and preparing edible roots and tubers for consumption, following safe digging and preparation techniques.

A Sobering Reminder:

Even for experienced foragers, mistakes can happen. The following cautionary points are crucial:

- Always Double-Check: Never rely solely on a single identification source. Consult multiple references and never hesitate to err on the side of caution if unsure about a plant.
- Beware of Look-Alikes: Many poisonous plants have close resemblances to edible species. Learn to distinguish look-alike plants carefully.
- When in Doubt, Throw it Out: If you have any doubts whatsoever about the safety of a foraged plant, discard it immediately. Your health is paramount.

Conclusion:

Foraging is a complex skill that requires extensive knowledge and experience. This chapter serves as a reminder of the inherent dangers involved and should not be a substitute for proper foraging education. If you are not a seasoned forager, prioritise obtaining reliable identification guides and seeking mentorship from experienced individuals before attempting to incorporate wild plants into your diet.

Chapter 6: First Aid and Basic Medical Care: Be Ready to Respond

Emergencies can strike unexpectedly, leaving you responsible for the health and well-being of yourself and others. This chapter empowers you to take charge by building a comprehensive first aid kit and equipping yourself with basic first-aid knowledge for common emergencies.

The Importance of First Aid:

First aid refers to the initial medical assistance provided to a person experiencing an injury or illness until professional medical help becomes available. Even simple first-aid measures can significantly improve the outcome of an emergency situation.

Building a Comprehensive First Aid Kit:

A well-stocked first-aid kit is an essential component of emergency preparedness. Here's a breakdown of essential supplies to consider including:

Basic Wound Care:

- Adhesive bandages: Assorted sizes of sterile adhesive bandages (adhesive strips and butterfly closures) for covering minor cuts, scrapes, and blisters.
- Non-stick sterile pads: Used for applying pressure to wounds and absorbing blood.
- Antiseptic wipes: For cleaning minor wounds and surrounding skin areas.
- Gauze pads and rolls: Sterile gauze pads in various sizes for dressing wounds and gauze rolls for creating larger bandages.

- Medical tape: Holds dressings and bandages firmly in place.

Pain Relief and Medication:

- Pain relievers: Over-the-counter pain relievers like acetaminophen or ibuprofen to manage pain and fever.
- Antihistamines: To reduce itching caused by insects and allergic responses.
- Antidiarrheal medication: For treating mild cases of diarrhoea.
- Personal medications: Include a sufficient supply of any prescription medications you or your family members rely on regularly.

Other Essential Supplies:

- Emergency blanket: Helps retain body heat in case of hypothermia.
- Instant cold compress: Soothes pain and reduces swelling from minor injuries.
- Thermometer: For monitoring body temperature.
- Disposable non-latex gloves: Protects you from bodily fluids while administering first aid.
- Scissors and tweezers: For removing splinters, ticks, or cutting bandages.
- Eye patch: Used to cover an injured eye and promote healing.
- Irrigation syringe (bulb syringe): For flushing wounds with clean water.
- First-aid manual: Provides a quick reference guide for performing basic first-aid procedures.

Tailoring Your First-Aid Kit:

- Family Needs: Consider any specific medical needs of family members when assembling your kit (e.g., asthma inhaler, allergy medication).
- Activity-Specific Needs: If you engage in specific activities like hiking or camping, include additional supplies relevant to those environments (e.g., tick removal kit, blister pads).
- Environmental Considerations: Adapt your kit based on your geographic location. For example, sunscreen and insect repellent may be essential in sunny or bug-infested areas.

Maintaining Your First-Aid Kit:

- Regular Inventory: Periodically check your kit and restock expired items or replace used supplies.
- Proper Storage: Store your kit in a cool, dry, and easily accessible location, ideally out of reach of children.

The Next Step: Basic First-Aid Knowledge

The following chapter will delve into basic first-aid knowledge for common emergencies, empowering you to respond effectively until medical help arrives.

First Aid and Basic Medical Care: Be Ready to Respond

Equipping yourself with a comprehensive first-aid kit is crucial, but preparedness extends beyond supplies. This chapter empowers you with basic first-aid knowledge for common injuries and illnesses, enabling you to respond confidently in emergency situations until professional medical help arrives.

Remember:

- Always prioritise safety: Ensure the scene is safe for yourself and the injured person before administering first aid.
- Seek immediate medical attention for serious injuries: If breathing is laboured, uncontrollable bleeding occurs, or a severe fracture is suspected, call emergency services or proceed to the nearest emergency room immediately.

Treating Common Injuries:

- Minor Cuts and Scrapes:
    - Clean the wound with gentle pressure and clean water.
    - Apply antiseptic wipes to disinfect the area around the wound.
    - Cover the wound with a sterile adhesive bandage.
- Bleeding:
    - Apply direct pressure to the wound with a sterile pad or clean cloth.
    - Elevate the injured limb if possible.
    - If bleeding persists after applying pressure for 10 minutes, seek medical attention.

- Sprains and Strains:
    - Apply the RICE principle: Rest, Ice, Compression, Elevation.
    - Apply ice wrapped in a cloth to the injured area for 15-20 minutes at a time, several times a day.
    - Use an elastic bandage to provide gentle compression (not too tight).
    - Elevate the injured limb to reduce swelling.
- Burns:
    - Cool minor burns with cool running water for 10-15 minutes.
    - Avoid putting ice directly on the burn.
    - Cover the burn with a sterile bandage loosely.
    - Seek medical attention for severe burns or burns that blister.

- Insect Bites and Stings:
    - Remove the stinger using tweezers if present, avoiding squeezing the venom sac.
    - Clean the area with soap and water.
    - Use a cold compress to ease the itching and swelling.
    - Apply calamine lotion or antihistamine cream to relieve itching.

Treating Common Illnesses:

- Fever:
    - Dress lightly and encourage hydration with cool liquids.
    - Use over-the-counter pain relievers like acetaminophen or ibuprofen following dosage instructions

(consult a healthcare professional for children or those with pre-existing medical conditions).

○ Seek medical attention for high fevers (over 103°F) or fevers that persist for more than 3 days.

- Diarrhoea:
    - Encourage clear fluids like water or broth to prevent dehydration.
    - Over-the-counter antidiarrheal medication can help manage symptoms.
    - Seek medical attention if diarrhoea is severe, bloody, or accompanied by high fever or vomiting.

- Vomiting:
    - Allow the person to vomit comfortably and avoid giving anything by mouth until vomiting subsides.
    - Once vomiting stops, offer small sips of clear fluids gradually.
    - Seek medical attention if vomiting is persistent, bloody, or accompanied by severe abdominal pain.
- Allergic Reactions:
    - If a person experiences a severe allergic reaction (anaphylaxis), identify and remove the allergen if possible.
    - If an EpiPen is available and the person has been prescribed one, administer the injection according to instructions.
    - Call emergency services immediately and monitor the person's breathing until help arrives.

Disclaimer:

The information provided in this chapter is for general knowledge purposes only and is not a substitute for professional medical advice. Always consult a qualified healthcare professional for diagnosis and treatment of any medical condition.

The Importance of CPR and First-Aid Training:

Formal first-aid and CPR training can significantly enhance your preparedness and equip you with the skills to respond effectively in life-threatening emergencies. Consider enrolling in a certified first-aid and CPR course to gain practical experience and build confidence in administering essential first-aid measures.

First Aid and Basic Medical Care: Be Ready to Respond (continued)

Wounds can be a common consequence of emergencies, and proper care is crucial to prevent infection and promote healing. This section highlights essential wound care and hygiene practices you can implement during emergencies to safeguard your health.

Understanding Wound Healing:

The human body has a remarkable ability to heal wounds. However, improper care can impede this process and increase the risk of infection. Here are the stages of wound healing:

- Inflammation: Immediately after an injury, the body initiates an inflammatory response, sending white blood cells to the area to fight infection. The symptoms of this stage include discomfort, swelling, and redness.

- Proliferation: The body starts rebuilding damaged tissue by generating new blood vessels and skin cells.
- Maturation and Remodelling: The newly formed tissue strengthens and matures over time, eventually regaining its original function.

Wound Care Principles to Prevent Infection:

- Stop the Bleeding: Apply direct pressure using a clean cloth or sterile bandage to control bleeding.
- Clean the Wound: Gently irrigate the wound with clean water to remove dirt and debris. Avoid using harsh soaps or antiseptics directly in the wound, as they can damage healing tissue.
- Debridement: For larger wounds, carefully remove any visible dirt or debris with sterile tweezers. If unsure about removing embedded objects, seek medical attention.

- Prevent Further Contamination: Cover the wound with a sterile adhesive bandage or dressing to protect it from dirt and bacteria.

Signs and Symptoms of Infection:

- Increased redness, swelling, and pain around the wound
- Pus drainage from the wound
- Fever
- Red streaks leading away from the wound

Maintaining Hygiene During Emergencies:

Maintaining proper hygiene during emergencies is equally important for preventing infections. Here are some key practices:

- Handwashing: Wash your hands thoroughly with soap and clean water for at least 20 seconds before and after cleaning a wound, or whenever they become visibly dirty.Use an alcohol-based hand sanitizer in the event that soap and water are not accessible.
- Bathing: Bathe regularly, even if with limited water resources. Prioritise cleaning the most critical areas like the face, hands, genitals, and underarms.
- Oral Hygiene: Brush your teeth and clean your mouth regularly, even if you don't have access to toothpaste. Use clean water or a homemade saline solution (mix 1/2 teaspoon salt with 1 cup of clean water).

Additional Tips:

- Monitor Wound Healing: Regularly observe your wound for signs of healing or infection.
- Change Dressings Regularly: Replace soiled or damp dressings to maintain a clean environment for wound healing.
- Promote Healing: A balanced diet and adequate sleep can contribute to faster wound healing.

By following these wound care and hygiene practices, you can significantly reduce the risk of infection and promote optimal healing during emergencies.

Chapter 7: Navigation and Communication When Technology Fails: Finding Your Way Without GPS

In today's world, dependence on technology for navigation is high. However, emergencies can disrupt communication and GPS signals. This chapter empowers you with essential land navigation skills using a map and compass, ensuring you can find your way even when technology fails.

The Importance of Map and Compass Navigation:

- Self-Reliance: Mastering map and compass navigation equips you with the ability to navigate independently, regardless of technological limitations.

- Improved Situational Awareness: Understanding your location relative to landmarks enhances your decision-making capabilities during emergencies.
- Adaptability: Map and compass skills empower you to navigate unfamiliar territory and unforeseen detours.

Essential Navigation Tools:

- Topographic Map: A topographic map provides detailed information about the terrain, including elevation contours, landmarks, and water sources. Choose a map with a scale appropriate for your needs.
- Base Plate Compass: A base plate compass features a rotating bezel with directional markings, an orienteering arrow, and a baseplate with straightedge and ruler for map measurements.

Map Reading Fundamentals:

- Map Symbols: Familiarise yourself with the legend on your map, understanding the symbols used to represent different features like roads, trails, bodies of water, and vegetation.
- Scale: The map scale indicates the relationship between the distance represented on the map and the actual ground distance. Learn to measure distances on the map using the scale and translate them to real-world distances.
- Contour Lines: On a map, contour lines show changes in height. Understanding how to read contour lines allows you to visualise the terrain and navigate effectively.

Compass Basics:

- Orienting the Compass: Hold the compass level in your hand with the baseplate facing you. The orienteering arrow should point towards your body.
- Magnetic North vs. True North: Be aware of the difference between magnetic north (indicated by the compass needle) and true north (grid north on most maps). Declination is the angle between these two directions and varies depending on your location. Account for declination when using your compass with a map.

Basic Land Navigation Techniques:

1. Triangulation: Identify your location on the map using at least two prominent landmarks visible from your current position.
2. Orient the Map: Hold the map flat and rotate it until north on the map aligns with the direction of magnetic north indicated by your compass needle (after accounting for declination).
3. Plan Your Route: Using your oriented map, identify your destination and plan your route by following trails, landmarks, or contour lines.
4. Taking Bearings: Point the compass in the direction you intend to travel. The bearing (degrees) displayed on the compass bezel corresponds to your direction of travel relative to magnetic north.

5. Maintaining Direction: While travelling, periodically check your compass to ensure you are following the intended bearing. Use landmarks on the map to confirm your progress.

Practice Makes Perfect:

Developing proficiency in map and compass navigation requires practice. Find a safe, familiar outdoor location and experiment with the techniques outlined above. As your confidence grows, gradually venture into more challenging terrain.

Additional Considerations:

- Pacing: Develop a consistent walking pace and estimate distances travelled based on the time spent walking.

- Dead Reckoning: Dead reckoning involves estimating your position based on your starting location, direction of travel, and distance travelled. This technique is less precise than using a compass but can be helpful in conjunction with map reading.
- Safety First: Always prioritise safety when navigating. Keep an eye on the weather, your surroundings, and any potential dangers.
- Seek Professional Instruction: Consider enrolling in a wilderness navigation course or seeking guidance from experienced navigators to enhance your skills and gain practical experience.

Navigation and Communication When Technology Fails (continued)

We've established the importance of map and compass navigation. This chapter delves into complementary techniques – utilising landmarks and celestial bodies – to empower you with a diverse navigational toolkit for emergencies when technology is unavailable.

Landmarks: Your Signposts in the Wild

Landmarks are prominent, fixed features in the landscape that serve as natural signposts. Here's how to leverage them effectively for navigation:

- Identify Key Landmarks: Train your eye to recognize distinct features like mountains, rivers, large rock formations, or unique vegetation patterns. The more distinctive a landmark, the easier it is to recall and utilise for navigation.

- Landmark Triangulation: Just as with a map, use at least two distinct landmarks visible from your current position. By drawing imaginary lines of sight between yourself and these landmarks on a map (if available), you can pinpoint your location. Even without a map, landmark triangulation helps confirm your general direction of travel.
- Landmark Relocating: Once you've identified and positioned yourself relative to landmarks, use them for continuous reference as you navigate. Memorise the order and relative positions of landmarks to retrace your steps if needed. For example, if you remember reaching a particular stream after passing a large, solitary rock, recognizing these landmarks in reverse order will guide you back to your starting point.

Celestial Navigation: A Glimpse at the Stars (For Advanced Users)

Celestial bodies offer an intriguing navigation method, but it requires more advanced skills and practice. Here's a basic introduction:

- Understanding Constellations: Familiarise yourself with recognizable constellations like Ursa Major (the Big Dipper) or Orion. These constellations maintain their relative positions throughout most of the night, allowing you to determine direction. Download a stargazing app or invest in a celestial navigation guide for comprehensive constellation identification.
- Polaris, Your Celestial Guide (Northern Hemisphere Only): Locate Polaris, the North Star. It appears nearly stationary near the North Pole.

By facing Polaris, you are facing north. This technique is invaluable for nighttime navigation, but remember, it's only applicable in the Northern Hemisphere.

Important Considerations:

- Weather Dependence: Both landmark and celestial navigation rely on clear visibility. Overcast skies or dense fog can render these techniques ineffective. Always have alternative navigation methods prepared in case of poor visibility.
- Time of Day: Celestial navigation becomes more challenging during daytime as the sun overpowers the visibility of most stars. Plan your navigation accordingly, prioritising landmark or map and compass techniques during daylight hours.

- Advanced Techniques: Advanced celestial navigation involves using a sextant to measure the angle of celestial bodies above the horizon. This requires specialised knowledge, practice, and nautical tables for calculations. Focus on mastering fundamental landmarks and basic celestial navigation before venturing into more advanced techniques.

Complementary Techniques for Enhanced Navigation:

- The Power of Combining Skills: Don't rely on a single navigation method. Combine map and compass with landmark and, if your skills allow, celestial navigation. This multi-layered approach provides redundancy and increases your resilience in unforeseen situations.

- Detailed Note-Taking: Maintain a detailed navigation log. Record landmarks observed, directions travelled, estimated distances covered, and even sketches of the landscape. These notes can be invaluable if you need to retrace your steps or communicate your location to rescuers.

Navigation and Communication When Technology Fails: Sending the Message When Lines Are Down

In today's world, communication often relies on cellular networks and internet connectivity. However, emergencies can disrupt these systems, leaving you isolated. This chapter explores alternative communication methods to stay connected during such situations.

The Importance of Alternative Communication:

- Maintaining Contact: Alternative communication methods empower you to maintain contact with loved ones or emergency services during technological outages.
- Search and Rescue: The ability to signal for help can be crucial in a life-or-death situation. Mastering basic communication techniques enhances the chances of a successful rescue.
- Community Building: In large-scale emergencies, establishing communication with others in your vicinity can foster collaboration and increase overall safety.

Basic Signalling Techniques:

- Visual Signalling: Use bright clothes, signal mirrors, or smoke signals to attract attention. Learn international distress signals like three large fires in a triangular pattern for ground-to-air communication.
- Audible Signalling: Utilise whistles, air horns, or banging on objects to create noise and attract attention. SOS distress signal (three short bursts, three long bursts, three short bursts) is universally recognized.

Communication with Specialized Equipment:

- Personal Locator Beacons (PLBs): These battery-powered devices transmit a distress signal to search and rescue satellites, providing your location for rescue efforts.

Register your PLB with the appropriate authorities for faster response.

- Ham Radio: Ham radio offers two-way communication capabilities over long distances without relying on cellular networks. A licence is required to operate a ham radio. Consider enrolling in licensing courses to gain the necessary knowledge and skills.

Limitations and Considerations:

- Line of Sight Restrictions: Visual and audible signalling methods often have limited range and may be ineffective in situations with obstructions or adverse weather conditions.
- Equipment Availability and Knowledge: PLBs and ham radios require investment and knowledge for proper operation

. Ensure you understand how to use these devices before relying on them in an emergency.

- Power Source Dependence: PLBs and some communication devices rely on batteries. Have a backup plan for recharging or replacing batteries to maintain communication capabilities.

Additional Considerations:

- Community Emergency Plans: Familiarise yourself with community emergency plans that may outline designated communication channels or meeting points during emergencies.
- Learning Sign Language: Basic sign language skills can facilitate communication with individuals who are deaf or hard of hearing, even in noisy environments.

Chapter 8: Security and Self-Defence Strategies: Protecting Yourself and Your Loved Ones

The safety and security of yourself and your loved ones are paramount. This chapter empowers you with strategies to create a secure home environment and deter potential threats.

The Importance of Home Security:

- Peace of Mind: Implementing effective home security measures provides peace of mind knowing you have taken proactive steps to deter crime and safeguard your belongings.
- Deterrence: A well-secured home discourages potential intruders by presenting a significant obstacle and increasing the risk of detection.

- Early Warning: Security systems can provide an early warning in case of an attempted break-in, allowing you to call for help and take necessary precautions.

Building a Layered Security Approach:

- Physical Barriers: Form the first line of defence. Strong doors and frames, reinforced windows with secure locks, and deadbolts on all exterior doors deter break-ins.
- Security Doors and Windows: Consider installing additional security features like reinforced door jambs, security window film, or window bars on vulnerable ground-floor windows.
- Door and Window Alarms: Install alarms on doors and windows to alert you and potentially scare off intruders if a breach occurs.

- Lighting: Adequate outdoor lighting deters nighttime activity around your home. Motion-sensor lights can further illuminate suspicious activity.
- Signage: Post visible security system warning signs to discourage potential intruders.

Additional Security Measures:

- Fencing and Gates: Secure your property perimeter with a fence and a locked gate, especially if you have a backyard.
- Landscaping: Trim bushes and trees around windows and doorways to eliminate hiding places for intruders.
- Valuables Inventory: Maintain an inventory of valuables with photos and serial numbers for insurance purposes and easier identification in case of theft.

- Secure Valuables: Store valuables in a safe or safety deposit box.

Maintaining Your Security System:

- Regular Testing: Test your security system and alarms regularly to ensure proper functioning.
- Battery Replacement: Replace batteries in smoke detectors and security system components promptly to avoid malfunctioning during an emergency.
- Home Alarms and Monitoring: Consider professional monitoring services for your security system for immediate response in case of an alarm trigger.

Situational Awareness: Your Key to Safety

Situational awareness is the ability to be aware of your surroundings and potential threats, allowing you to make informed decisions and react proactively to avoid danger. Here's how to cultivate this essential skill:

- Be Present and Engaged: Avoid distractions like phones or music while walking or in public spaces. Pay attention to your surroundings using all your senses: sight, sound, and even intuition.
- Identify Potential Hazards: Look for suspicious activity, unfamiliar individuals, poorly lit areas, or anything that makes you feel uneasy. Trust your gut instinct.
- Maintain Safe Distances: Avoid isolated areas and keep a safe distance from people exhibiting erratic behaviour.

- Plan Escape Routes: When entering a new environment, mentally identify potential exits and escape routes in case of an emergency.

Personal Safety Habits for Everyday Life

- Travel Smart: Choose well-lit, populated routes when walking alone, especially at night. Tell someone where you plan to go and when you expect to arrive. Avoid using shortcuts through deserted areas.
- Be Mindful of Your Belongings: Keep bags and purses close to your body, preferably in front. Avoid carrying large sums of cash or displaying valuables in public.
- Project Confidence: Walk with purpose and confidence. Maintain eye contact with those around you but avoid appearing confrontational.

- Trust Your Gut: If something doesn't feel right, it most likely is. Don't be afraid to remove yourself from an uncomfortable environment or politely decline unwanted interactions.
- Be Vocal: If you are harassed or feel threatened, don't hesitate to speak up firmly and attract attention. Shout for help or use a personal safety alarm if necessary.

Empowering Bystanders:

Bystanders can significantly impact a situation. Here's how you can contribute to a safer environment:

- Be an Observer: Pay attention to your surroundings and be alert to potential problems.
- Report Suspicious Activity: If you witness something concerning, report it to the authorities promptly.

- Offer Help: If someone appears lost or in need of assistance, offer help in a safe and public manner.

Non-Lethal Self-Defense Techniques and De-Escalation Strategies

De-Escalation: The First Line of Defense

Before resorting to physical techniques, de-escalation should always be your primary strategy. Here are some key tactics:

- Maintain Distance: This creates a physical barrier and gives you time to react.
- Use Calming Communication: Speak slowly and clearly, avoid accusatory language, and focus on diffusing the situation.
- Identify Exits: Always be aware of your surroundings and plan escape routes in case de-escalation fails.

- Non-Verbal Communication: Maintain eye contact, stand tall with an open posture, and avoid fidgeting to project confidence.

Non-Lethal Techniques: For When De-Escalation Fails

Important Note: While I can provide an overview, it's crucial to seek professional training to master these techniques safely and effectively.

- Target Vulnerable Areas: Aim for strikes that can temporarily disable an attacker, such as the groyne, knees, or solar plexus.
- Evasive Manoeuvres: Practice footwork and dodging techniques to create space and avoid getting hit.
- Use of Blocks: Learn effective blocks to deflect punches, kicks, and grabs.

Advanced Techniques (For Trained Individuals Only):

- Joint Locks and Pain Compliance Techniques: These require precise knowledge of anatomy and proper application to avoid injury.
- Takedowns: Techniques to bring an attacker to the ground for control, but should only be attempted with proper training.

Remember:

- Non-lethal techniques are not foolproof. Their effectiveness depends on the situation and the attacker's size and aggression.
- The goal is to escape, not engage in a fight. Use these techniques only to create an opportunity to get away safely.

- Be aware of your surroundings. Don't use techniques that could put yourself or others in further danger (e.g., tripping near stairs).

Legal Considerations

Self-defence laws vary by region. Familiarise yourself with the legal justifications for using force in your area.

Seek Professional Training

For your safety, it's critical to enrol in a reputable self-defence program taught by qualified instructors. They can provide hands-on training, ensure proper form to avoid injury, and help you build the confidence needed to react effectively in a dangerous situation.

Part 3: Mastering Long-Term Preparedness

Building on the foundation laid in the previous sections, Part 3 dives deeper into strategies for long-term preparedness. This goes beyond just emergency kits and sheltering in place – it's about cultivating resilience and adaptability in the face of unforeseen challenges.

Key Areas of Focus:

- Self-Sufficiency Skills:
    - Food Production and Preservation: Explore gardening techniques, food storage methods, and alternative food sources.

- Water Acquisition and Purification: Learn methods for collecting rainwater, purifying water from natural sources, and water conservation.
- Energy Production: Investigate alternative energy options like solar panels, wind turbines, and biofuels.
- Basic Repairs and Maintenance: Develop skills for fixing clothing, tools, and other essential equipment.
- Community Building and Collaboration:
  - Strengthening Social Networks: Build strong relationships with neighbours and like-minded individuals who share your preparedness goals.
  - Barter and Exchange Systems: Explore alternative methods of acquiring goods and services beyond traditional currency in case of economic collapse.

- Shared Resource Management: Develop plans for collaborating on tasks like food production, security, and infrastructure maintenance within your community.
- Advanced Risk Assessment and Planning:
    - Scenario Planning: Consider a wider range of potential threats beyond natural disasters, such as economic collapse, pandemics, or social unrest.
    - Long-term Sheltering Options: Explore alternative sheltering options beyond your home, considering factors like sustainability and defensibility.
    - Bug Out Bag (BOB) Evolution: Refine your BOB to accommodate longer-term needs, including tools, medical supplies, and communication equipment.

Developing a Long-Term Mindset:

Part 3 emphasises the importance of cultivating a preparedness mindset. This includes:

- Resilience Training: Practise mental exercises to build emotional fortitude and a positive outlook in challenging situations.
- Adaptability: Be prepared to adjust your plans and strategies as circumstances evolve.
- Lifelong Learning: Continuously seek new knowledge and skills to enhance your preparedness.

Chapter 9: Staying Informed and Monitoring Threats: Knowledge is Power

In an emergency, staying informed about the situation is crucial for making sound decisions and ensuring your safety. This chapter focuses on identifying reliable news sources during such times, when misinformation can run rampant.

The Challenge of Misinformation:

Emergencies create a breeding ground for rumours, fake news, and sensationalised reports. These can lead to panic, hinder response efforts, and put people at risk.

Strategies for Identifying Reliable Sources:

- Reputable News Organisations: Prioritise established news outlets with a history of accurate reporting. Look for news sources that fact-check their information and have a clear distinction between news and opinion pieces.
- Government Sources: Official government websites and social media pages from relevant agencies (e.g., emergency management, public health) are reliable sources for updates and instructions.
- Local Media: Local news outlets often have a deeper understanding of the specific situation in your area and can provide targeted information.

Verification Techniques:

- Check for Author Credentials: Look for articles written by journalists with expertise in the relevant field.
- Cross-Reference Information: Don't rely on a single source. Verify information by comparing reports from multiple reputable outlets.
- Be Wary of Sensationalized Headlines and Emotional Appeals: Sensationalized language is often used to grab attention, not provide accurate information.
- Investigate the Source: Check the website's "About Us" section to understand their background and editorial policies. Be wary of unfamiliar websites with no clear ownership or editorial process.

- Look for Evidence: Reputable sources will cite their sources and provide evidence to support their claims.

Utilising Technology for Information Gathering:

- Emergency Alert Systems: Sign up for official emergency alert systems in your area to receive real-time updates.
- Verified Social Media Accounts: Follow verified social media accounts of reliable news organisations and government agencies. However, be cautious of unverified accounts spreading rumours.

Understanding Local Emergency Alert Systems

Chapter 9 emphasises the importance of staying informed during emergencies, and local emergency alert systems play a crucial role in achieving that. This section dives into how these systems work and how you can leverage them to your advantage.

What are Local Emergency Alert Systems?

Local emergency alert systems are communication networks designed to notify residents about critical situations in their immediate area. These alerts can be issued through various channels, including:

- Television and Radio Broadcasts: Emergency Alert System (EAS) broadcasts interrupt regular programming to deliver essential information during emergencies.

- Wireless Emergency Alerts (WEA): Compatible mobile devices receive text-like messages with details about the threat and instructions on how to stay safe.
- Landline Phone Calls: Automated systems can deliver pre-recorded emergency messages to landline phones in affected areas.
- Local Authority Websites and Social Media: Government websites and social media pages of relevant agencies (e.g., emergency management) often provide updates and instructions during emergencies.

Benefits of Local Emergency Alert Systems:

- Timely Warnings: These systems provide immediate notification of threats, allowing you to take necessary precautions.
- Clear Instructions: Alerts typically include information about the nature of the emergency and recommended actions to stay safe.
- Widespread Reach: They can reach a large population quickly and efficiently, ensuring everyone is aware of the situation.

Enrolling in Local Emergency Alert Systems (if applicable):

- Contact your local emergency management office. They can provide information on how to sign up for WEA alerts in your area (if available) and advise on other local notification methods.

- Register for alerts on local authority websites and social media pages.

Important Considerations:

- Not all emergencies trigger alerts. The decision to activate the system depends on the severity and scope of the threat.
- System limitations may exist. Technical malfunctions or limitations in coverage areas can sometimes occur.

Building Strong Community Networks for Mutual Support

In times of crisis or even everyday challenges, a strong community network can be an invaluable source of support. These networks go beyond just having friendly neighbours; they're about fostering genuine connections and building a web of reciprocity where everyone contributes and benefits. Here's how you can actively participate in building such a network:

Identify Your Needs and Assets:

- Start by reflecting on what you can offer and what you might need. Are you handy with repairs? Do you enjoy cooking for others? Perhaps you're a great listener or have organisational skills.

Recognizing your strengths allows you to contribute effectively.

- Similarly, consider your needs. Would you appreciate help with childcare? Maybe you need someone to watch your dog while you travel. Identifying your needs helps you connect with people who can offer that support.

Seek Out Existing Networks:

- Look for local community groups, social clubs, or online forums that align with your interests or needs. These groups can provide a platform to connect with like-minded people and offer opportunities for collaboration.
- Venture beyond your comfort zone without fear. Explore groups focused on activities you've always wanted to try. You might discover hidden talents and forge unexpected friendships.

Become a Good Neighbour:

- Introduce yourself to your neighbours, even if it's just a friendly wave or a short conversation. Simple gestures can pave the way for stronger relationships.
- When an opportunity arises, lend a helpful hand. Maybe it's helping someone carry groceries or offering to watch their kids while they run errands. Small acts of kindness go a long way in building trust and fostering a sense of community.

Initiate Activities and Gatherings:

- Don't hold off on taking the initiative until someone else does. Organise potlucks, game nights, or book clubs in your neighbourhood. These events provide a relaxed space for people to connect and build friendships.

- Focus on shared interests. Organise activities around your passions, whether it's gardening, board games, or book clubs.

Practice Open Communication and Empathy:

- Listen well and genuinely express interest in the lives of others around you. Building trust and understanding is key to a strong support network.
- Offer help and support without expecting anything in return. The spirit of reciprocity will naturally grow in a supportive community.

Utilise Technology for Connection:

- Create a neighbourhood online group or forum. This can be a space to share information, resources, and offer support even when you can't physically meet.

- Use social media platforms to connect with local groups and events. Many communities have dedicated online spaces for communication and collaboration.

Chapter 10: Mental and Physical Preparedness: Building Resilience - Managing Stress and Anxiety in Uncertain Times

Emergencies and challenging situations can trigger significant stress and anxiety. Chapter 10 emphasises the importance of mental and physical preparedness to navigate these uncertainties effectively. This section focuses on strategies for managing stress and anxiety during difficult times.

Understanding Stress and Anxiety:

- Stress: The body's natural response to a perceived threat or challenge. It can manifest physically (increased heart rate, muscle tension) and emotionally (irritability, worry).

- Anxiety: A feeling of unease, worry, or nervousness about a future event or outcome. It can be a normal reaction to uncertainty but can become overwhelming if not managed effectively.

The Impact of Stress and Anxiety on Preparedness:

- Impaired Decision-Making: Stress and anxiety can cloud judgement and hinder your ability to make clear, rational decisions in critical situations.
- Reduced Coping Ability: Feeling overwhelmed can make it difficult to take necessary actions to prepare or respond effectively.
- Physical Health Consequences: Chronic stress can lead to headaches, digestive issues, and a weakened immune system.

Strategies for Managing Stress and Anxiety:

- Develop Healthy Habits: Prioritise regular exercise, a balanced diet, and adequate sleep. These habits promote overall well-being and bolster your resilience during stressful times.
- Relaxation Techniques: Practise relaxation techniques like deep breathing, meditation, or mindfulness exercises to calm your mind and body.
- Maintain a Positive Outlook: Focus on what you can control and visualise positive outcomes. Gratitude exercises can also shift your perspective and reduce anxiety.
- Limit Media Consumption: Constant exposure to negative news can exacerbate anxiety. Seek updates from reliable sources, but limit your overall consumption.

- Connect with Your Support Network: Talk to trusted friends, family members, or a therapist about your anxieties. Sharing your burdens can lighten the load and gain valuable perspectives.
- Develop Coping Mechanisms: Identify healthy activities that help you manage stress, such as spending time in nature, listening to music, or pursuing hobbies.

Building Resilience:

- Develop a Sense of Control: Take charge of what you can control, such as your preparedness efforts and how you react to situations.
- Practice Acceptance: Accept that some things are beyond your control and focus your energy on what you can influence.

- Learn from Challenges: View challenges as opportunities to learn and grow. Analyse situations to improve your preparedness for the future.

Maintaining Physical Fitness and Overall Health: The Bedrock of Preparedness

A strong foundation of physical fitness and overall health is crucial for preparedness in any situation. Chapter 10 likely expands on this concept, but here's a general overview:

Physical Fitness:

- Regular Exercise: Try to get in at least 150 minutes a week of moderate-to-intense aerobic activity or 75 minutes a week of strenuous activity.

This strengthens your cardiovascular system, improves endurance, and helps manage weight.

- Strength Training: Incorporate strength training exercises that target all major muscle groups at least twice a week. This builds muscle mass, improves bone density, and enhances functional abilities for everyday tasks.
- Flexibility: Regularly stretch to maintain good range of motion and improve overall flexibility. This helps prevent injuries and improves overall well-being.

Overall Health:

- Eat a well-balanced diet that is high in whole grains, fruits, vegetables, lean protein, and other nutrients. This provides your body with the essential nutrients it needs to function optimally and fight off illness.

- Hydration: Make sure you are getting enough water throughout the day. Performance both mentally and physically can suffer from dehydration.
- Good Sleep: Try to get 7-8 hours of good sleep every night. Sleep is essential for physical and mental recovery, cognitive function, and immune system health.
- Preventative Care: Make time for routine examinations with your dentist and physician.Early detection and treatment of potential health issues can prevent complications and improve your overall well-being.

Benefits of Maintaining Physical Fitness and Overall Health:

- Increased Resilience: A healthy body is better equipped to handle stress and resist illness.
- Improved Physical Performance: You'll have greater strength, stamina, and agility, which can be crucial in emergency situations.
- Enhanced Mental Well-being: Physical activity releases endorphins, which have mood-boosting effects and can help combat anxiety and depression.
- Stronger Immune System: A healthy lifestyle bolsters your immune system's ability to fight off infections and diseases.

Remember:

- Small changes add up. Start with small, achievable goals and gradually increase the intensity and duration of your workouts and healthy habits over time.
- Find activities you enjoy. Exercise shouldn't feel like a chore. Choose activities you find fun and engaging to ensure long-term consistency.
- Consult your doctor before starting any new exercise program, especially if you have any pre-existing health conditions.

By prioritising physical fitness and overall health, you're laying the groundwork for a resilient and adaptable body that can handle the challenges that may arise. This investment in your well-being is a cornerstone of effective preparedness for any situation.

Cultivating a Positive Mindset and Adaptability: Essential Tools for Preparedness

When faced with uncertainty and challenges, a positive mindset and adaptability become invaluable assets. Chapter 10 likely delves deeper, but here's a breakdown of how to cultivate these crucial aspects of preparedness:

Developing a Positive Mindset:

- Focus on the Controllable: Shift your focus from what you can't control to what you can. This empowers you to take action and navigate situations effectively.
- Practice Gratitude: Regularly acknowledge the positive aspects of your life, big or small. Gratitude fosters a sense of optimism and resilience.

- Visualisation: Visualise yourself successfully handling challenging situations. This positive mental rehearsal boosts confidence and reduces anxiety.
- Positive Self-Talk: Challenge negative thoughts and replace them with encouraging self-affirmations.

Enhancing Adaptability:

- Embrace Change: Recognize that change is inevitable. Instead of resisting it, view it as an opportunity for growth and learning.
- Open-mindedness: Be open to new ideas and approaches. This broadens your perspective and allows you to adapt to unforeseen situations.
- Problem-Solving Skills: Develop your problem-solving skills by actively seeking solutions and learning from past experiences.

- Resourcefulness: Train yourself to think creatively and utilise available resources to overcome challenges.

The Synergy Between Positive Mindset and Adaptability:

A positive mindset fosters the belief that you can overcome challenges, while adaptability equips you with the skills to do so. This powerful combination is key to navigating difficult situations effectively.

- Positive thinking fuels your motivation and perseverance when faced with obstacles.
- Adaptability allows you to adjust your plans and strategies as circumstances evolve.

Remember:

- Cultivating a positive mindset and adaptability is an ongoing process. Practise these skills daily to strengthen your mental resilience.
- Learn from setbacks. View challenges as opportunities to learn and grow, making you more adaptable in the future.
- Surround yourself with positive people. The company you keep can significantly influence your outlook and approach to challenges.

By fostering a positive mindset and developing adaptability, you equip yourself with powerful tools to navigate uncertainty and emerge stronger from challenging situations. This combination is a cornerstone of effective preparedness for anything life throws your way.

Chapter 11: Sharpening Your Skills and Refining Your Plan - Conducting Regular Drills for Evacuation Scenarios and First Aid

Building a preparedness plan is essential, but Chapter 11 emphasises that it's just the first step. This chapter likely focuses on the importance of practising your plan and honing your skills through regular drills. Here's a breakdown of key areas to practise:

Evacuation Drills:

- Practice Makes Perfect: Regular evacuation drills familiarise everyone in your household (or community) with escape routes, designated meeting points, and procedures in case of fire, natural disaster, or other emergencies.

- Different Scenarios: Conduct drills for various scenarios. Practice nighttime evacuations, simulating situations where exits might be blocked, and account for individuals with mobility limitations.
- Time is of the Essence: Track your evacuation time during drills. Aim to improve efficiency with each practice run.
- Include Your Pets: If you have pets, incorporate them into your evacuation drills. Familiarise them with their carriers and practice transporting them calmly.

First Aid Drills:

- Learn Basic Techniques: Ensure everyone in your household (or community) has basic first aid knowledge. This could include CPR, wound care, and how to handle common emergencies like choking or allergic reactions.

- Hands-on Practice: Don't just memorise steps – practice applying first aid techniques on mannequins or willing participants.
- Keep Supplies Handy: Ensure your first aid kit is well-stocked and easily accessible during drills. Practice locating and using the supplies you might need in a real emergency.
- Review Regularly: First aid knowledge and CPR certification can expire. Schedule regular refresher courses to keep your skills sharp.

Additional Considerations:

- Debrief After Drills: After each drill, discuss what went well and identify areas for improvement. This helps refine your plan and address any potential shortcomings.

- Involve Everyone: Include all household members (or community members) in drills. This empowers everyone to take an active role in their preparedness.
- Make it Fun: Gamify drills, especially for children. This can make practising essential skills more engaging and less stressful.

Remember:

- Regular drills are not about achieving perfection; they're about building muscle memory and confidence in your ability to respond effectively during an emergency.
- The more you practise, the more likely you are to react calmly and efficiently in a real-world situation.
- Drills are also opportunities to identify gaps in your plan or areas where you need additional training or supplies.

By incorporating regular drills for evacuation scenarios and first aid into your preparedness routine, you can significantly enhance your skills and refine your plan. This proactive approach can make a world of difference when responding to an emergency.

Maintaining and Upgrading Your Prepping Supplies: Ensuring Readiness Over Time

Having a well-stocked prepper pantry and emergency kit is crucial, but their effectiveness hinges on proper maintenance and thoughtful upgrades. This ensures your supplies remain functional and relevant to your evolving needs. Here's a guide to keeping your preps in top shape

Regular Inventory and Inspection:

- Schedule Checkups: Set regular reminders (quarterly or biannually) to thoroughly inspect your supplies.
- Rotate Stock: Pay attention to expiration dates on food, water, and medications. Rotate them into your regular consumption to prevent waste and ensure you have readily available supplies in case of an emergency.

- Check Functionality: Test batteries, flashlights, fire extinguishers, and other equipment regularly. Replace dead batteries and ensure everything functions as intended.
- Physical Inspection: Look for signs of damage or wear and tear on items like tents, backpacks, or first aid supplies. Repair or replace damaged items promptly.

Upgrading Your Preps:

- Evolving Needs: As your life circumstances change (growing family, dietary restrictions, new hobbies), your preparedness needs may evolve. Review your supplies regularly and adjust accordingly.
- New Technologies: The world of preparedness is constantly evolving.

Research new tools and technologies that might enhance your preparedness strategy. Consider incorporating them into your existing plan.

- Lessons Learned: Real-world events or personal experiences can highlight gaps in your preparedness. Use these lessons to identify areas for improvement and upgrade your supplies to address them.
- Multi-Purpose Items: Look for versatile items that can serve multiple functions. This reduces overall storage space and simplifies your prepping inventory.

Storage Best Practices:

- Organisation is Key: Clearly label and organise your supplies for easy access in an emergency.
- Climate Control: Store food and medications in cool, dry locations to prevent spoilage or degradation.
- Moisture Control: Use moisture absorbers or airtight containers to prevent moisture damage to essential documents, electronics, or first aid supplies.
- Pest Control: Take measures to prevent pests from contaminating your food or other supplies.

Remember:

- Maintenance is an ongoing process, not a one-time event. Regularly checking and updating your supplies ensures they remain reliable when you need them most.
- Don't be afraid to adapt. Your preparedness journey is fluid, so adjust your supplies and plan as your needs and circumstances change.
- Quality over quantity. Invest in high-quality, durable items that will last.

By following these tips, you can maintain and upgrade your prepping supplies, ensuring they remain a reliable safety net in times of need. Remember, a well-maintained prep is a proactive approach to safeguarding yourself and your loved ones.

Continuously Learning New Skills to Stay Ahead of the Curve: A Pillar of Preparedness

In today's rapidly changing world, complacency is the enemy of preparedness. Chapter 9 likely emphasises the importance of staying informed, but this section dives deeper into the concept of lifelong learning as a cornerstone of preparedness. Here's why continuously acquiring new skills is essential:

Evolving Threats and Challenges:

- The Unexpected: New threats and challenges can emerge unexpectedly. Equipping yourself with a diverse skill set allows you to adapt and respond effectively to unforeseen situations.
- Technological Advancements: The world is constantly evolving, and preparedness strategies need to keep pace.

Learning new technologies can enhance your ability to communicate, access information, and manage resources during emergencies.

- Shifting Needs: As your life progresses, your preparedness needs might change. Mastering new skills can ensure you're prepared for different scenarios, whether it's raising children, caring for elderly parents, or navigating a changing job market.

Benefits of Lifelong Learning:

- Enhanced Problem-Solving Skills: Learning new things strengthens your critical thinking and problem-solving abilities, valuable assets in any challenging situation.
- Increased Confidence: Mastering new skills boosts your confidence and self-reliance, empowering you to handle unexpected situations.

- Improved Adaptability: A diverse skill set allows you to adapt to changing circumstances and find creative solutions to unforeseen problems.
- Personal Growth: Lifelong learning is a journey of self-discovery and personal growth. It keeps your mind sharp and opens doors to new possibilities.

Strategies for Continuous Learning:

- Formal Education: Consider taking courses, workshops, or attending seminars to acquire new skills or knowledge relevant to preparedness.
- Online Resources: The internet is a treasure trove of free and paid learning resources. Explore online courses, tutorials, and educational websites.
- Books and Audiobooks: Reading books on various preparedness topics is a convenient and enriching way to expand your knowledge base.

- Hands-on Learning: Learning by doing is often the most effective method. Seek opportunities to practise new skills through volunteering, attending workshops, or participating in practical exercises.
- Mentorship: Find a mentor with expertise in areas you'd like to learn more about. They can provide guidance, share their knowledge, and offer valuable insights.

Remember:

- Lifelong learning is a marathon, not a sprint. Set realistic goals and celebrate small achievements along the way.
- Learning should be enjoyable. Choose topics that interest you to maintain motivation and make the process more engaging.

- Focus on developing a well-rounded skill set. Don't limit yourself to just survival skills. Consider learning communication skills, financial literacy, or basic engineering principles – all of which can contribute to your overall preparedness.

By embracing lifelong learning, you equip yourself with the adaptability and resourcefulness needed to navigate the uncertainties of life. This commitment to continuous improvement strengthens your preparedness foundation and empowers you to face challenges with confidence.

Conclusion: The Empowered Prepper: Ready for Anything - The Importance of Long-Term Planning and Staying Vigilant

The Empowered Journey:

This book has guided you through a comprehensive exploration of preparedness, empowering you to become a proactive agent in safeguarding yourself and your loved ones. You've learned to navigate emergencies with a calm mind, a well-stocked pantry, and a diverse skill set.

The Power of Long-Term Planning:

Effective preparedness isn't about a one-time event; it's a continuous journey. By developing a long-term plan, you establish a roadmap that evolves alongside your life's circumstances and the ever-changing world around you.

Staying Vigilant:

Complacency is the enemy of preparedness. Stay informed about potential threats, emerging technologies, and best practices in the preparedness community. Continuously seek new knowledge and refine your plan to stay ahead of the curve.

Beyond Emergencies:

The skills and knowledge you've acquired extend beyond emergency preparedness. Resourcefulness, adaptability, and a commitment to self-reliance are valuable assets in everyday life.

Remember:

- Preparedness is a mindset, not just a checklist. Cultivate a proactive approach to challenges and embrace the ongoing process of learning and improvement.
- Empower your community. Share your knowledge and skills with others. Building a strong and supportive network strengthens overall preparedness.
- Don't be afraid to adapt. The world around us is constantly changing, so be prepared to adjust your plan and strategies as needed.

Peace of Mind Through Preparedness: A Lifelong Journey

Living in an uncertain world can be unsettling. Natural disasters, economic disruptions, or even personal challenges can cause anxiety and a sense of vulnerability. However, there's a powerful antidote to this unease: preparedness.

Preparation is not about fear-mongering. It's about taking proactive steps to build resilience and navigate challenges with confidence. This book has equipped you with the knowledge and tools to embark on a lifelong journey of preparedness, ultimately leading to a greater sense of peace of mind.

The Pillars of Preparedness:

- Building a Foundation: This involves creating a comprehensive emergency plan, assembling essential supplies, and establishing a communication strategy.
- Developing Self-Sufficiency: Learning basic skills like gardening, food preservation, and minor repairs fosters independence and adaptability.
- Building a Strong Community: Connecting with neighbours and like-minded individuals creates a support network for resource sharing, collaboration, and emotional well-being.
- Maintaining Physical and Mental Fitness: A healthy body and mind are essential for effectively handling stressful situations.

- Lifelong Learning: The world of preparedness is constantly evolving. Continuously seeking new knowledge and skills keeps you ahead of the curve.

The Journey is the Reward:

Preparedness isn't a destination; it's a continuous odyssey of learning, adapting, and refining your approach. Embrace the process – it empowers you, builds confidence, and fosters a sense of control amidst uncertainty.

Benefits Beyond Emergencies:

The skills and knowledge you gain extend far beyond emergency preparedness. Resourcefulness, problem-solving abilities, and a commitment to self-reliance are valuable assets in everyday life, enhancing your ability to navigate personal challenges and unexpected situations.

Remember:

- Start small and build momentum. Don't feel overwhelmed by the vastness of preparedness. Begin with manageable steps and gradually build upon your foundation.
- Focus on progress, not perfection. There's no single "right" way to be prepared. Tailor your plan to your specific needs and circumstances.
- Celebrate your achievements. Acknowledge your progress, no matter how small. This reinforces positive habits and keeps you motivated on your preparedness journey.

Peace of Mind is Within Reach:

By embracing preparedness as a lifelong pursuit, you're actively cultivating peace of mind. Knowing you have the skills, resources, and support network to face challenges empowers you to live life with greater confidence and security. So, take the first step today, and embark on your empowering journey towards a prepared and peaceful future.

Bonus Appendix:

Master Prepper Checklist: A Comprehensive List of Essential Supplies (Printable)

Disclaimer: This checklist is a starting point and can be customised based on your specific needs, location, and threats.

Emergency Food and Water:

- Non-perishable food (2-week supply for each person)
- Water purification tablets or water filter
- Water storage containers (enough for at least 3 days per person)
- Can opener (manual)

Sanitation and Hygiene:

- Toiletries (soap, shampoo, toothpaste, feminine hygiene products)
- Sanitation wipes
- Toilet paper (enough for at least 1 week per person)
- Trash bags
- Disposable plates, cups, and utensils

Shelter:

- Tent (if evacuation is necessary)
- Tarp or emergency shelter
- Sleeping bags or blankets (enough for everyone)
- First-aid kit (comprehensive with essential supplies)
- Prescription medications (extra supply)
- Cash (small bills)

Lighting and Communication:

- Flashlights (multiple, with extra batteries)
- Battery-powered radio (NOAA weather radio recommended)
- Whistle or signalling device
- Cell phone charger (portable if possible)

Tools and Equipment:

- Multipurpose tool
- Duct tape
- Work gloves
- Dust mask
- Fire extinguisher
- Local maps
- Compass

Clothing and Bedding:

- Sturdy shoes or boots
- Rain gear
- Warm clothes (appropriate for your climate)
- Change of clothes for each person
- Extra blankets

Optional Items:

- Fire starter kit
- Bartering items (e.g., cigarettes, small denomination cash)
- Entertainment (books, cards, board games)
- Pet food and supplies (if applicable)
- Important documents (copies of passports, IDs, insurance documents) stored in a waterproof container

Remember:

- Regularly rotate your food and water supplies to ensure freshness.
- Check and replace batteries in flashlights and other equipment periodically.
- Update your first-aid kit and medications as needed.
- Tailor this list to your specific needs and potential threats in your area.

This checklist is a printable resource to help you get started on gathering your essential preparedness supplies. Remember, this is not an exhaustive list, and you may need to adjust it based on your unique circumstances. By using this checklist as a starting point and conducting further research, you can ensure you have the supplies you need to weather any emergency.

Glossary of Prepping Terms: Understanding the Language of Preparedness

This glossary defines key terms you might encounter in the world of preparedness, referencing the items mentioned in the Master Prepper Checklist (Appendix):

- Bartering Items: Goods or small valuables used for trade in situations where traditional currency might be less useful (e.g., cigarettes, lighters, small denomination cash).
- Bug-Out-Bag (BOB): A backpack or other container filled with essential supplies for evacuation scenarios (mentioned under Shelter in the Appendix).
- Can Opener (Manual): A non-electric can opener to access canned food during emergencies (mentioned under Emergency Food and Water).

- Cash (Small Bills): Having a small amount of cash readily available can be helpful in emergencies, especially if electronic transactions are unavailable (mentioned under Shelter).
- Duct Tape: A versatile adhesive tape with a wide range of uses in repairs, securing items, or creating makeshift shelters (mentioned under Tools and Equipment).
- Dust Mask: A mask to protect your respiratory system from dust particles or debris during emergencies (mentioned under Tools and Equipment).
- Emergency Food and Water: A stockpile of non-perishable food (enough for at least 2 weeks) and water (enough for at least 3 days per person) to sustain you in an emergency (mentioned in a separate category).

- Emergency Shelter: A temporary shelter option like a tent or tarp used during evacuation or when a permanent dwelling is compromised (mentioned under Shelter).
- First-Aid Kit: A portable kit containing medical supplies to treat minor injuries and illnesses (mentioned under Shelter).
- Flashlights (Multiple, with Extra Batteries): A reliable source of light during power outages or nighttime emergencies. Carrying extra batteries ensures continued use (mentioned under Lighting and Communication).
- Local Maps: Physical maps of your area can be invaluable if GPS or electronic navigation becomes unavailable (mentioned under Tools and Equipment).
- Master Prepper Checklist: A comprehensive list of essential supplies for preparedness, referenced in this glossary (mentioned in the title).

- Multipurpose Tool: A single tool with various functionalities (e.g., pliers, knife opener, screwdriver) for addressing different repair or survival needs (mentioned under Tools and Equipment).
- NOAA Weather Radio: A battery-powered radio that receives National Oceanic and Atmospheric Administration (NOAA) weather alerts and emergency broadcasts (mentioned under Lighting and Communication).
- Non-Perishable Food: Food items with a long shelf life that don't require refrigeration and can be safely consumed after an extended period (mentioned under Emergency Food and Water).
- Sanitation and Hygiene: Supplies like soap, toilet paper, and sanitation wipes to maintain cleanliness and hygiene during emergencies (mentioned in a separate category).

- Sleeping Bags or Blankets: Essential for staying warm during an evacuation or in compromised shelter situations (mentioned under Shelter).
- Shelter: Supplies and equipment to create temporary housing during an evacuation or when your primary residence is unusable (mentioned in a separate category).
- Tarp: A large sheet of waterproof material used for creating a makeshift shelter, collecting rainwater, or signalling for help (mentioned under Emergency Shelter).
- Tools and Equipment: Items that can be used for repairs, building shelters, or addressing various survival needs (mentioned in a separate category).
- Water Purification Tablets or Water Filter: Methods to purify water from potentially unsafe sources to make it drinkable (mentioned under Emergency Food and Water).

- Water Storage Containers: Containers for storing water for emergency use (mentioned under Emergency Food and Water).
- Whistle or Signalling Device: A way to attract attention and signal for help in emergency situations (mentioned under Lighting and Communication).

Sample Emergency Plans: Customizable Templates for Family and Evacuation Planning

Disclaimer: These templates are starting points and should be adapted to your specific family needs, dwelling situation, and potential threats in your area.

Part 1: Family Emergency Plan

This plan outlines your family's response strategy for various emergencies.

General Information:

- Family Names and Contact Information: List all family members' names, ages, and contact information (work, cell phone, etc.)
- Meeting Location: Designate a meeting location inside your home for emergencies that don't require evacuation (e.g., basement, interior hallway).

- Out-of-Town Contact: Identify an out-of-town friend or relative as a family contact point in case you're separated during an emergency.

Communication Plan:

- Communication Methods: Establish how you'll communicate with each other during an emergency if phone lines are down (e.g., texting, designated radio channel).
- Out-of-Town Contact Role: Assign the out-of-town contact as a central information hub for family members to check in during emergencies.

Shelter Plan:

- Evacuation Decision-Making: Discuss the criteria for deciding to evacuate (e.g., approaching fire, mandatory evacuation orders).
- Evacuation Route: Plan your evacuation route, considering traffic patterns and potential road closures.
- Bug-Out-Bag (BOB) Contents: Review the items in your family's BOBs (refer to Master Prepper Checklist in Appendix) and ensure everyone knows where they're stored.

Shelter in Place Plan:

- Shelter Location: Identify a safe location inside your home to shelter in place during emergencies like tornadoes or hazardous material spills (e.g., basement, interior room with no windows).

Additional Considerations:

- Pet Plan: Include a plan for sheltering or evacuating pets with adequate supplies (food, water, leash, etc.).
- Special Needs: Address any specific needs of family members (medical conditions, mobility limitations) in your emergency plans.
- Practice Drills: Conduct regular practice drills to familiarise everyone with the plan and ensure effective response during emergencies.

Part 2: Evacuation Plan

This plan focuses specifically on the evacuation process.

Evacuation Triggers:

- List the specific situations that would trigger an evacuation (e.g., natural disasters, chemical spills).

Evacuation Tasks:

- Outline a step-by-step process for evacuating your home efficiently (e.g., turn off utilities, shut windows, gather essential supplies).
- Assign specific tasks to family members to ensure everyone contributes to the evacuation process.

Transportation:

- Designate a primary and secondary evacuation vehicle and ensure they are well-maintained and fueled.
- Consider alternative transportation options in case vehicles are unavailable (e.g., walking, biking).

Route and Destination:

- Identify your evacuation route, considering traffic patterns and potential road closures.
- Determine your evacuation destination (relative's house, shelter location) and have a backup location in case your primary destination is inaccessible.

Communication:

- Reiterate your communication plan from the Family Emergency Plan (refer to Part 1).
- Ensure everyone has a charged cell phone and access to a portable charger if possible.

Remember:

- Tailor these templates to your specific needs and discuss the plans with all family members.
- Regularly review and update your plans as your family circumstances or environmental threats change.
- Practice evacuation drills to ensure a smooth and coordinated response during an emergency.

By creating and practising these emergency plans, you can significantly improve your family's preparedness and ability to respond effectively in various emergency situations.

www.ingramcontent.com/pod-product-compliance
Lightning Source LLC
Chambersburg PA
CBHW071206240526
45470CB00018B/1525